INTRODUCTION TO ENTERPRISE RISK MANAGEMENT

N. Krishnamurthy

Safety Consultant and Trainer
www.profkrishna.com
Singapore

DEDICATION

To my students and colleagues who have inspired and challenged me to do better.

ISBN: 1539436284
ISBN-13: 978-1539436287

INTRODUCTION TO ENTERPRISE RISK MANAGEMENT
N. Krishnamurthy

TABLE OF CONTENTS

———

PREFACE

This is a book for beginners. It is written for the diploma or degree student, and the practicing engineers, particularly in small and medium enterprises (SMEs). However author hopes that even the risk management veteran may find something of value in these pages.

With certain exceptions, it is well known stuff, except the way it is presented, and possibly in viewpoint and interpretation in certain matters.

In 2007, author wrote his first book on the subject titled *"Introduction to Risk Management"* and published it himself in print. It was aimed at a Singapore audience mainly because the small island nation in which he had settled down after exciting professional experience in in USA, India, and Singapore, with teaching and research in universities and training and consulting for industry in structural engineering, computer applications, workplace safety and risk management, had just then made a heavy revamp of its regulations with its Workplace and Safety Act of 2006.

When the first print sold out the author started revising it for a second edition. After discovering the popularity of Amazon Self-Publishing books, author decided to revise the book for a global audience with minimal changes in other areas, and release it fast as an E-book. He also chose a fresh title: *"Introduction to Enterprise Risk Management"*, to distinguish it from most risk management books already available for the financial sector.

He is also publishing this as an E-book through Kindle.

During the intervening years, no earth-shaking developments have taken place in the subject, but the author has become more extensively and intensively involved in workplace safety and risk management, developing and teaching more courses, presenting more talks and publishing more papers, conducting personal and sponsored research, and training and consulting for the government and private parties on related topics.

He has patented a computer-based invention on risk management in Singapore and Australia, titled 'The SAFER Diamond'. More details of his activities and access to his publications may be obtained from author's website: www.profkrishna.com

This book retains almost all the material from the earlier book – including his *"5A-Way to Safety"*. However, many sections have been re-organized, and a lot of fresh material has been added to elucidate difficult concepts and expand the scope of applications.

Author retains Singapore as a role model because Singapore has shot up from a developing nation to a developed one in a few decades, and its safety record has improved by leaps and bounds in the last decade. That is why many of the regulations and case studies presented in the book reflect author's Singapore experience.

However, having lived in USA for long, and with his familiarity and interest in the practice of industrial risk management in other countries, author is aware that risk management practices vary widely around the world.

To globalize this edition, author has provided basic references in Chapter 3 to give readers access to relevant information on risk management in other parts of the world.

Author has tried to present the basics in a simple, chatty and story-telling style that he adopts in his lectures, avoiding bombast, and (he hopes!) boredom. His light touch should not be mistaken for flippancy.

Author admits that certain ideas he has presented herein may be new or unusual, and some may even deviate from conventional wisdom or published literature. Readers may choose to agree or disagree with his views – and certainly adopt alternative approaches, where available.

Author also seeks the reader's indulgence when he uses 'I' and other first person pronouns, especially when he relates personal anecdotes.

He invites readers to visit his website [*http://www.profkrishna.com*] both to access certain additional information and resources on safety-related matters, and also to provide him feedback and suggestions on the contents of the book for corrections and improvements. His e-mail is: proscank@gmail.com

While acknowledging the use of extensive existing resources in forming and refining his ideas, the author regrets he is not able to make specific citations from this wealth of specialized knowledge in every case.

The author has drawn freely on the experience of his predecessors and contemporaries, as well as from the wisdom of the ages as applicable. While tapping the fountainhead of information that Internet offers in the public domain, he has been careful not to use copyrighted material without attribution, to the best of his knowledge. Any contravention would be be unintentional.

The few typos, omissions and inconsistencies that the author missed during the proof reading of the first printing have been rectified in this reprint.

January 2017   *N. Krishnamurthy*

———

1. OPEN YOUR THIRD EYE!

First, a word about human awareness and pro-active attitudes.

I often start my lecture on hazard identification by showing the audience a poster on safety, and asking: *"How many of you have seen this before?"* [*See* left part of Fig. 1.1.]

Fig. 1.1 - "We look ... but do not see!"

Generally there is no response, even when I repeat the question. Occasionally a single hand goes up half-heartedly, and I ask *"Where?"*

He (or she) will say, *"In a MOM seminar talk"* or something like that. *[MOM stands for Singapore's Ministry of Manpower. – NK]*

I dramatically point to the double doors at the back of the room, through which all the participants had come in, exited for the coffee break and come in again, a total of three times – and ask the entire group to turn around. [*See* right part of Fig. 1.1.]

And there, stuck on both the doors, at eye level, is this

very same poster! [Fig. 1.2.]

Fig. 1.2 - Violation Report to MOM

The sad truth is: We look ... but do not see! We listen ... but do not hear! We touch ... but do not feel!

Then I tell them my philosophy for risk management (and lots of other things!):

All of us must learn to see, hear, smell, taste, touch – in short, sense everything better, all around us, above and below us.

Why? Because dangers surround us, waiting to pounce upon us, or insidiously work upon us, to trap us, maim us, kill us.

[I use the term 'we' only out of courtesy – I trained myself for this a long time ago! – Author]

Having shocked them into attention, I go ahead and complete my pitch:

I tell them that the poster they saw (or didn't see!) was MOM's message for encouraging people to report any safety violations they happen to see.

A universal problem is that most incidents – "near misses" that escaped becoming accidents, not due to any quick evasive action by the personnel involved, but by a lucky break – and even many minor accidents – do not get recorded or reported.

Laziness, 'loss of face', fear of criticism or punishment, and lack of motivation are some of the reasons for not reporting these incidents.

Now, as a new initiative and part of the WSH Act, 'Incident Reporting' regulations have been instituted, according to which, a dangerous occurrence, defined as a serious workplace incident in which no one is killed or injured, <u>must</u> be reported to MOM in a specified format.

At top left of the poster (Fig. 1.2) it says: *"At work today, you noticed the scaffolding has come loose and may collapse at any time. You are worried someone may fall."*

Whether they had 'noticed' or they were 'worried' until now, they will/must notice and must be worried from now on.

Then I go over the three numbered points lower down in the poster:

1. What do I do?

"Tell your supervisor or your project manager. He will do the rest."

Whether they had been telling the project manager or not and he* had been doing the rest or not, now my class knows that from now on they must, and he must because it is an offence not to report, and not to act upon it.

*[*NOTE: Use of the male pronoun shall equally refer to the female counterpart also, unless obviously gender-specific. – Author]*

2. If they won't help me, what can I do?

Call 6317 1111" [in Singapore]

I reassure them that they do not have to give their name, phone number, or IC Number!"

3. Can I keep quiet and do nothing?

No! By keeping quiet and doing nothing, an accident may happen. And the victim could be you."

I add an extra twist: *"Or it could be somebody you know ... ore even somebody you don't know! – Would it, should it, matter?"*

I close with the comment: *"See, now your third – intuitive – eye is open!"*

A participant adds: *"The third ear also?"*

Yes, plus a sixth sense too.

———

2. WHY BOTHER?

2.1. BACKGROUND

Before we get into the subject of risk management we must ask why we must bother about it at all, and why the authorities are so seriously concerned and taking such concerted action on it.

At the top of the list must appear the occupational injury experience of the industry and the country in the matter of workplace safety. One of the convenient measures is the 'Fatality Rate' namely the number of deaths for every 100,000 (full-time) workers.

This number can vary from a little less than 1 (– no nation has zero workplace fatalities) to the high 30s or beyond, Where a nation has not reported its fatality rate, it is taken as 11.0, as some kind of average of high fatality rate countries.

Figure 2.1 displays the five-year average fatality rates for 2011-2015 sourced from the International Labour Organization website.

2.2. RISK MANAGEMENT AS MISHAP PREDICTOR

A primary reason for a nation – or an industry or organization – to consider risk management is that it is one of the few leading indicators (that is, signs that predict) what can go wrong in a project and how we can plan to avoid or manage it.

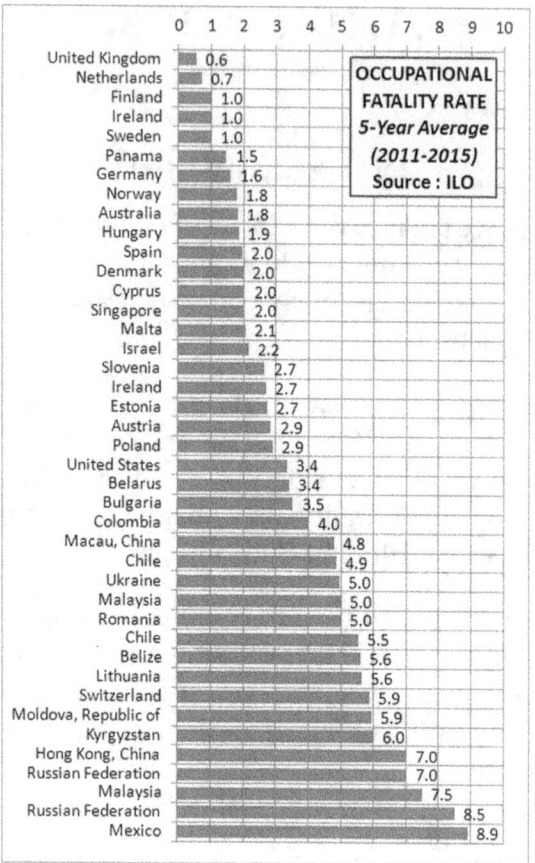

Fig. 2.1. - Fatality Rate in various countries (2011-2015)

Accident statistics indicate that most accidents are triggered or caused by an activity or condition not noted earlier. While rarely such a cause might have been unknown or not experienced earlier, in the majority of cases the cause simply had not been identified, due to lack of or inadequate risk assessment.

So when the workplace accidents and injuries go up, or are already higher than the controlling authority likes, risk assessment and as a natural corollary risk control should be the first recourse.

For example, Singapore, being a small, well organized nation with a disciplined citizenry, already had a decent fatality rate in the early 2000s. It was 18th in a list of more than 100 countries, with a workplace fatality rate of 4.9 in 2004. Yet it aspired to go higher so as to become attractive to regional and global investors.

Three high-profile accidents of 2004 (Nicoll Highway, Fusionpolis, and Keppel, details of which may be found from author's website) also created around the world a negative image of Singapore's industry safety record requiring prompt corrective measures.

Finally, as seen from Fig. 2.2. (sourced from MOM) since 2002 the accident statistics seemed to have reached a plateau at 2.2 accidents per million hours worked, implying that existing safety measures by themselves might not lead to any further improvement.

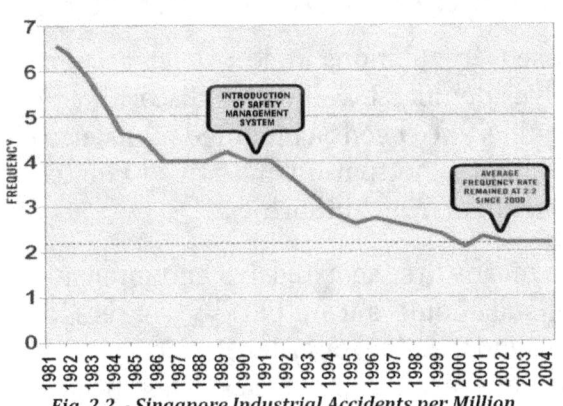

Fig. 2.2. - *Singapore Industrial Accidents per Million Man-Hours Worked (1981-2004)*

Note in the figure another plateau a few years earlier had been rejuvenated by a specific action namely the introduction of a Safety Management System.

Out of such challenges arise the search for new initiatives to reduce the fatality rate and improve overall workplace safety. Codes of Practice provide a strong motivating force.

Singapore responded with the introduction of WSH Act in March 2006 to replace the time-worn Factories Act, and set the goal of reducing the fatality rate to 2.5 per 100,000 by the year 2015.

[It reached this goal in 2008, so it raised the bar to reach 1.8 in 2018. It reached this goal too in 2015, but has fallen a little to about 2.0 in 2016.]

The Act incorporated risk management regulations in September 2016 as noted in Chapter 3.

2.3. WORKPLACE SAFETY AS CORE VALUE

Revised regulations and stricter penalties by themselves do not sit well with a discerning population. Proper adherence needs a modified mindset, a change of heart, kind of a 'bottom-up' initiative to match the 'top-down' mandate from authorities.

This means first an extensive and intensive dialogue with all stakeholders from CEO to workers, conducting seminars and training for all levels, focused at identifying hazards in various industries and trades, and at assessing and controlling consequent risks, combined with providing resources and incentives.

(a) The Singapore Response:

Apart from Singapore Contractors Association Limited (SCAL) and Institution of Engineers (Singapore) Academy (IESA) –with whose efforts author has been closely associated – other organizations also conduct seminars and short courses, with MOM endorsement.

Slow but steady progress has been made on this front. The message went out loud and clear:

- Safety starts with risk assessment; safety needs risk management.

- Do a risk assessment, regardless of specialty or size.

- Do it before a project starts.

- Do it immediately for ongoing projects.

- RA/RM reports need not be submitted to the authorities. But companies must have complete documentation ready, if and when the inspectors come ... even without an accident!

- Tell yourself, remind yourself: *"I accept workplace safety as a core value ... like I accept myself, my family, my community, my country as core values."*

(b) General guidelines:

There are no prescriptive rules for all of this. But there are numerous guidelines and performance requirements.

- For routine situations do not wait for, or depend upon, the authorities to tell you what to do. Do it yourself. You know your problems best. For special or very critical cases, they will guide.

- Do not wait for accidents to happen. Anticipate,

9

prevent them.

- Do not ignore incidents ("near misses"). Document and report them, for the general good.

- Spread the risk, share the risk, transfer the risk.

- Share your problems and solutions with your cohorts. That will be the win-win situation.

- Communicate, across, up and down. Communicate clearly. Document your communication. The importance of proper communication cannot be under-estimated. – *See* news-item at end of the chapter!

The author likes to add the following layers to this thinking:

- Prepare yourself for the long haul: Benefits of investment in safety will:
  1. Take time to bear fruit, like the maturing of a tree; and
  2. Be spread around the industry, and may not be just one to one on the individual company investments – again like a tree may drop its fruit into a neighbor's yard and vice versa.

  - Think of the workers and staff as your workplace family. If *n* workers walk into your site every morning; *n* workers must go home safe and sound every evening – on their own two feet.

- Safety is not a 9 to 5, Monday to Friday thing. It should become a "24/7" (24 hours a day, 7 days a week) awareness and commitment, be it at home or workplace, on street or vacation.

- It should be a matter of pride that the industry is taking responsibility for its own safety rather than wait to be told. In advanced countries, industry tells the authorities what is needed.

2.4. SAFETY CULTURE

All the preceding should serve to define the oft-repeated but little understood phrase, 'safety culture'. There are scores of accepted definitions of the phrase, but author offers the following as his distilled essence version:

"Safety culture is what we think, say, and do, to maintain and promote the safety and welfare of our fellow human beings, other living things and natural resources, and the environment."

Safety culture is not a fad or a fashion. Evidence of safety culture may be simple acts such as the following:

- Covering up clothes-drying pole holder pipes when not in use;

- Picking up a banana peel in your path and dropping it into a dust bin rather than just stepping around it – possibly saving a pregnant wife or a doddering grandfather a few steps behind you from slipping on it;

- Taking seriously posters and videos, recommendations and warnings, that authorities provide for safety – and act on them;

- Reporting any missing scaffold planks at a worksite, unattended baggage at MRTs, and violations of safety norms anywhere;

[*"Have you?"* you may ask. Yes, I have reported unattended baggage a few times – each time officials have taken it seriously and checked it out, without asking for any personal information. It turned out that in these cases, investigation cleared the suspicion of threat.]

- Learning more about what can cause harm, who can be hurt, and how every one of us can help all of us avoid it;

- Thinking of our employees like 'family' where safety is concerned, in the sense that we are responsible for their sustenance, welfare, progress, and safety outside our homes as much as we are for our actual families' safety inside our homes; and,

- Making a habit of seeing better, hearing better, feeling better, understanding better – to catch signs of danger.

Safety culture is also getting up the courage, and acting promptly:

- To stop young kids from jumping around on escalators;

- To stop workers from using cell-phones when on the job; and,

- To warn workers, the very <u>first</u> time they ignore the PPE rules, that they will be dismissed if they are caught doing it again.

Safety culture is thus rectifying a situation which can cause harm to others, if not to ourselves. In short, safety culture is concern about the safety of others as much as our own. It should become a way of life.

In the final analysis, safety culture should translate to company loyalty and national pride. Not just the workforce and the management of an industry should be responsible for workplace safety, but the entire public must be involved.

It boils down to 'duty of care', common humanity that we must take care of the persons who are exposed to harm and loss for our sake, and on our behalf.

Author suggests employers asking themselves, *"Would I want my son or daughter to do this task under these conditions?"*

The National Safety Council of America advocates its workers to take this pledge:

"I pledge to:
- *Never compromise my own safety or the safety of my co-workers to get the job done*
- *Actively look for hazards, promptly report them, and take appropriate action to warn others*
- *Be a good safety role model for my friends and family even when off the job."*

That is why we must bother!

2.5. BENEFITS OF RISK ASSESSMENT

Apart from professional, legal, social, and humane reasons, there are many inherent advantages in increasing safety through RA and RM, such as the following:

- Presently in Singapore, risk assessment has become the prime factor for authorities to judge the compliance of companies with the WSH Act, and to examine as the root cause for accidents. Hence conducting RA will become an essential component of safe practice.

- Company branding and bidding criteria will include safety records, which will be based on RA and RM.

- RM will form part of safety management systems, to a much greater extent than heretofore.

- Customers will seek out companies with good safety records, again automatically endorsing and promoting regular RA.

- RA naturally leads to development of safe work practices (SWPs).

- RA will be a technique for self-assessment of a company's performance, and even staff evaluation.

- Accumulated risk register records can lead to elimination of bad practices and to innovation and improvement of good practices.

- Beyond all these should be the 'Business Case' for safety, namely, RA and RM are critical to safety, and safety is good business.

2.6. RISK MANAGEMENT IN PRACTICE – 1

singapore news · weekend TODAY · January 13 - 14, 2007

Firm fined $30,000 for death of two workers

Deficient risk assessment:

Investigations also showed [Company] merely devised a safety check list for [Sub-contractor], which in turn, only gave out oral instructions for workers to check if the structure was secure before using it.

An adequate hazard analysis would have shown that *"a visual inspection would not suffice"*, the investigation concluded.

———

3. SAFETY CODES AND REGULATIONS

Almost every country has its own industrial safety regulations. They may go by different names such as Code of Practice, Directive, Regulation, Guideline, Advisory, etc., ranging from a legal imperative to a recommendation.

But all of them will be aimed to improve the personal health and safety of the workers as well as the safety record and culture of the company and industry.

Most regulations on workplace safety would govern the health and welfare of personnel. However, after this primary need has been met, other factors must be addressed.

These would include, not necessarily in order of importance, property loss, environmental degradation, time delay, reputation damage, the applicable legal violations, and finally the all-pervasive financial impact.

Assessment of occurrences of mishaps and control of their consequences from such multiple factors in an industry would raise the scope of analysis and action to 'Enterprise Risk Management'.

We will track the practice of risk management in certain countries.

3.1. RISK MANAGEMENT IN UK

UK has the longest and most extensive involvement in the conventional risk management area. Some useful references are given below:

1. *Risk assessment – A brief guide to controlling risks in the workplace*

   http://www.hse.gov.uk/risk/controlling-risks.htm

2. *Risk Management* [This site offers many interactive e-tools, Examples and other materials for risk management. – NK]

   http://www.hse.gov.uk/risk/

3. *Reducing Risks – Protecting People* [On 'ALARP' meaning "As Low As Reasonably Practicable"– NK]

   http://www.hse.gov.uk/risk/theory/r2p2.pdf

3.2. RISK MANAGEMENT IN USA

USA's Occupational Safety and Health Administration (OSHA), established in 1970s has very detailed regulations and recommendations for a vast number of specific tasks and industries, but not a common code of practice for risk management for all industries and workplaces. Many states also have detailed guidelines. A lot of useful material is available on these topics from the Internet.

One general guideline and a few specific documents are given below:

1. *Job Hazard Analysis, OSHA 3071, 2002(Revised)*

   https://www.osha.gov/Publications/osha3071.pdf

2. *Risk Assessment – How do we know what we should be working on?* [Slide presentation by an OSHA grantee – NK]

   https://www.osha.gov/dte/grant materials/fy11/s h-22246-11/RiskAssessment.ppt

3. *Example – Application of 1910.119(e)(3)(vii)* [Highly hazardous chemicals]

   https://www.osha.gov/OshDoc/Interp pdf/I20050 201A appendix.pdf

3.3. RISK MANAGEMENT IN EUROPEAN UNION

As an organization, the European Union is quite young relative to UK and USA, but because of its size and the versatility of its member nations, it has developed a lot of information, knowledge, and digital tools in risk management.

The following are good sources for further information:

1. *Guidance on risk assessment at work (Directive 89/391/EEC)*

   https://osha.europa.eu/en/legislation/guidelines/g uidance-on-risk-assessment-at-work

2. *OiRA: free and simple tools for a straightforward risk assessment process* – and related publications

   https://osha.europa.eu/en/tools-and-publications/oira

3.4. RISK MANAGEMENT IN AUSTRALIA

Australia and many of its states have extensive regulations and detailed guidelines on risk management in general and applications to specific industries. The following should provide ample resources.

1. *Risk assessment and management*

   http://www.industry.gov.au/resource/Documents/ LPSDP/LPSDP-RiskHandbook.pdf

2. *AS/NZS ISO 31000:2009:Risk management – Principles and guidelines* [The standards are a priced publication, but the coverage may be viewed from:}

   http://www.finance.gov.au/sites/default/files/COV 216905 Risk Management Fact Sheet FA3 23082 010 0.pdf

3.5. RISK MANAGEMENT IN SINGAPORE

Singapore introduced the Workplace and Safety Act in March 2006, and the Risk Management Regulations in Sept. 2006. It legislated the Code of Practice for Risk Management in 2011, which has been revised twice.

1. *Singapore Workplace Safety and Health Act (Chapter 354A) –March 2006*

   http://statutes.agc.gov.sg/aol/search/display/view. w3p;query=DocId%3Aa7b4b808-d195-44ec-aa3d-dd5b1fa938f3%20Depth%3A0%20Status%3Ainfor ce;rec=0;whole=yes

2. *WSH ACT 2006 (Act 7 of 2006) – Workplace Safety And*

Health (Risk Management) Regulations 2006

https://www.wshc.sg/files/wshc/upload/cms/file/2014/WORKPLACE%20SAFETY%20AND%20HEALTH%20_RISK%20MANAGEMENT_%20REGULATIONS%202006.pdf

3. *Code of Practice on Workplace Safety and Health (WSH) Risk Management*

   https://wshc.sg/files/wshc/upload/cms/file/CodeOfPractice_RiskManagement_SecondRevision.pdf

4. A vision for the future:

   https://www.wshc.sg/files/wshc/upload/cms/file/Improving_WSH_Management_Singapore.pdf

3.6. MORE ON THE SINGAPORE SCENE

As already mentioned, Singapore has focused and invested a lot of resources into workplace safety and risk management in recent years. Ranking eighteenth worldwide in safety in 2004, it has arrived within the top five in the last couple of years.

Needless to say, similar events have occurred and actions have been taken in other developed and developing nations even earlier.

But Singapore may be taken as a showcase for these efforts for developing nations, and as a review case study for advanced nations.

(a) The Workplace Safety and Health (WSH) Act:

In Singapore, the push for risk management for all

and by all came with the Singapore Workplace Safety and Health (WSH) Act, to replace the older Factories Act, with effect from 1 March 2006.

To quote from the Act, it is *"an essential part of a new framework to cultivate good safety habits in all individuals at the workplace – from top management to the last worker. It requires every person at the workplace to take reasonably practicable steps to ensure the safety and health of every workplace and worker."*

Clause 47 of the Act states thus on what is "reasonably practicable" in the above, as follows:

"Where in any proceedings for an offence under any provision in this Act, it is alleged that any person failed to comply with a duty to do something so far as is reasonably practicable, it shall be for the accused to prove that — (a) it was not reasonably practicable to do more than what was in fact done to satisfy that duty; or (b) there was no better practicable means than was in fact used to satisfy that duty."

In other words, the question after any accident would be: Could you have done any better with the resources you had? Common practice and knowledge in the relevant industry must be considered in this decision.

In the first phase, the Act covered the three high-risk industries namely construction worksites, manufacturing establishments, and shipyards. In 2008, six more sectors were added, and in 2011, all the remaining workplaces were brought into the net, to protect all workers in Singapore.

(b) Three aims of the Act:

- To improve the workplace safety and health (WSH) standards

- To achieve reduction in fatality rate by half in 2015

- To attain WSH standards comparable to European countries

(c) Three key principles of the Act:

- To reduce risk at source by requiring all stakeholders to eliminate or minimize the risks they create at the workplace

- To instill greater industry ownership of occupational safety and health (OSH) standards. Focus will be shifted from complying with prescriptive requirements, to making employers responsible for developing work and safety procedures suited to their particular situations in order to achieve desired safety outcomes.

- To prevent accidents through higher penalties for poor safety and health management.

(d) Three key features of the Act:

- Liabilities for a range of persons at the workplace instead of focusing on the occupier

- Focus on effective management of workplace safety and health to achieve a safe outcome instead of prescribing rules

- Greater penalties for compromising safety and health

(e) Persons who have duties and responsibilities:

The following are the main stakeholders in the project:

- Employer
- Principal
- Occupier
- Contractor
- Manufacturer or supplier
- Erector or installer
- Worker
- Self-employed Person

The employer and principal have the main responsibility to conduct risk assessments to remove or control risks to workers at the workplace.

'Principal' is defined as one who engages another person to supply labor or do any work for compensation.

'Occupier' is the person who holds the certificate of registration. In all other workplaces, the occupier is the person who has control of the premises, regardless of whether they are the owner of those premises.

(f) Penalties:

The following are samples of certain penalties imposed under the Act.

- Individuals violating the Act, maximum fine SG$200,000 and/or maximum of 2 years imprisonment for first offence, twice the fine for subsequent offences.

- Corporations violating the Act, maximum fine SG$500,000 for first offense, and twice the fine for subsequent offences.

- Individual failing to comply with remedial order,

maximum fine SG$50,000 and/or maximum of 2 months imprisonment; additional fine of SG$5,000 for each day the offence continues.

- Individual failing to comply with stop work order, maximum fine SG$500,000 and/or maximum of 12 months imprisonment; additional fine of SG$20,000 for each day the offence continues.

Penalties in Singapore are not as high as in many Western countries. Being a small and disciplined nation, the enforcement actions in Singapore are quite thorough and swift.

More and more citations of violations and court cases are highlighting the absence or inadequacy of risk assessment as a main cause of accidents. [*See* 'Risk Management in Practice - 2' at end of Chapter 3.]

(g) WSH (Risk Management) Regulations 2006:

These regulations of the Act came into effect on 1 September 2006. According to these regulations, in every workplace, the employer, self-employed person and principal shall conduct a risk assessment in relation to the safety and health risks posed to any person who may be affected by his undertaking in the workplace.

He shall further take steps to eliminate or control the risks, maintain records, disseminate relevant information, and review or revise the assessment as needed.

Contravention of these requirements will lead to the liability of:

- For a first offence, a fine not exceeding $10,000; and

- For a second or subsequent offence, a fine not exceeding $20,000 or imprisonment for a term not exceeding 6 months or to both.

(h) Other initiatives of Ministry of Manpower:

Singapore's Ministry of Manpower (MOM) has a vastly expanded website with extensive Internet resources on workplace safety, including the following:

- A weekly *'OSHAlert'* (now changed to *'WSHAlert'*) bulletin, providing details of most accidents pointing out 'Do-s' and 'Don't-s'.

- The new Incident Reporting Regulations have already been mentioned.

- MOM has initiated ProBE (Programme-Based Engagement), as part of its Strategic Occupational Safety and Health Engagement Framework, to raise standards in areas which MOM has identified as priority or high-risk activity such as scaffolding and metal working.

- MOM has set aside a $5 million Risk Management Assistance Fund (RMAF) to help small and medium enterprises defray cost of engaging consultants to conduct RA and build up their in-house OSH capability.

- The Workplace Safety and Health Council (WSHC) was formed in 2008 to develop and disseminate safety information. In 2011 the Workplace Safety and Health Institute (WSHI) was established to conduct research.

- A program instituted by WSHC is BizSAFE, s five-tier benchmark of training and track record for Small and

Medium size Enterprises (SMEs).

Details of these and other initiatives may be found in MOM and WSHC websites, namely www.mom.gov.sg and www.wshc.sg.

3.7. AUTHOR'S WEBSITE

Author's papers on risk management and related subjects may be accessed from his 'Publications' and 'Downloads' pages of his website: www.profkrishna.com

While not claiming to be exhaustive, author has collected and tried to present relevant material that interested him from public domain, and made available his own publications and comments for the use of visitors to his site.

The site also includes information on his academic and professional background, anecdotes and links to other valuable resources from around the world.

For those who would like to borrow any of author's material, permission is granted for fair academic and research use of material from his website on condition that such use is acknowledged with the citation: *"Sourced from Dr. N. Krishnamurthy's website: www.profkrishna.com."* If it is a publication, do cite the publication in a standard format.

3.8. RISK MANAGEMENT IN PRACTICE – 2

Escalator Mess-up!

This was the up-escalator into the immigration area at a famous airport in S.E. Asia in the 1980s. (Replaced by a very huge, modern facility now. So don't worry!). I was there. I saw the whole thing.

Passengers who had arrived, tired but excited about visiting this famous place, had got on the escalator, and were being whisked up to the landing, shown in Fig. 3.1.

Fig. 3.1. - Escalator to Immigration

See the closed door at the top? It was locked – from the other side. Now, place yourself on the escalator, and imagine what was happening every second.

How many can the landing hold, even crushed against each other?

Before long, the landing was jam packed, men shouted, women screamed, children cried. Still the escalator kept bringing up live humans as if they were toys on a mass production conveyor belt.

At the last moment, before people spilled over the sides of the landing and crashed to the floor, the attendant who was to have opened the door a few minutes before the passengers disembarked from the plane, arrived and opened the door. He had been delayed at the toilet.

Don't ask about the red stop button. Maybe there was no button at that time. Escalators were new then. We didn't know enough about them.

How would you handle this situation, if you did a risk assessment today?

Lessons learned:

This problem could and should have been anticipated even before emergency buttons and risk management became important! Many deficiencies and corrective measures are obvious:

1. The switch to turn on the escalator should be under the control of the attendant in charge of opening the door so he could open the door and then turn on the escalator.

2. A second attendant should have been positioned at the bottom to receive and guide the landing passengers, making sure the door was open before the escalator was started.

3. The landing could have been sufficiently large to hold one plane-load of passengers.

4. The immigration door could have been of clear glass, so that:

    (a) someone on the other side could have opened the door, or,

    (b) an enterprising passenger on the landing could have broken the glass and turned the handle to open the door – unless the door was locked by a key – also with the top attendant in the toilet!

This is an object lesson in how risk assessment could be a leading indicator of accidents. Once the problem had been identified, there would be more than one solution to solve it!

———

4. HAZARD AND RISK

4.1. HAZARD

'Hazard' is simply potential danger, not danger itself. Electricity for instance is potentially very dangerous, but we use it casually all the time, as it is controlled so well, and we are so familiar with its safe use.

Hazard may be a place (desert, ocean), a situation (smog, intense cold), an animate entity (lion, criminal), an inanimate entity (scaffold, stored chemicals), energy source (sun, cooking gas), natural phenomena (lightning, earthquake), or an actual event (accident, collapse).

When hazards arise from events, an activity serves as the trigger mechanism. A damaged or improperly erected ladder may stand up by itself, but when a user climbs on it, it may collapse – in which case 'Climbing on the ladder' will be the activity and 'Falling from height' will be the hazard.

Even when the triggers are not events, some activity will usually be associated with the escalation of a hazard into risk, and can be identified as the trigger. Thus, while fire is a common hazard to people and property, the activity that will serve as a trigger for an accident is when a child or an arsonist sets it off in ignorance or for some evil gain.

That is why most risk assessment forms start off with the activity or job step for the associated hazard identification, and the entry must have a verb to indicate the trigger action. Job step or activity cannot be a mere noun such as 'Chemical' or 'Scaffold'. It has to be 'Mixing chemicals', 'Dismantling scaffold' etc. before they can be taken as causing hazards.

Hazards may be 'active' as when a flood or a fire spreads and engulfs its victims, or 'passive' as when a live electrical wire has lost its insulation and is waiting for a victim to touch it.

Hazards may be physical (injury), chemical, electrical, mechanical, environmental, ergonomic (working posture), biological, psychological, nuclear, financial, etc.

Hazards exist all around us, although we don't recognize most of them – and most of what we do recognize, we don't worry about!

- The foods we eat have 'expiry dates' because they are processed to last only up to a certain period. Many processed foods have a number of pesticides, preservatives, coloring agents etc. which, if their limits are exceeded, can cause health problems.

- The air we breathe in towns is laden with the exhaust fumes from the many vehicles which run on diesel or petrol, and from the many factories that produce essential goods. Singapore has good pollution control, but often it gets large doses of smog blown in from neighboring countries.

- The ozone layer blanketing our earth has been punctured, thanks to excessive use of fossil fuels by certain "advanced" countries, and everybody around the world is getting a larger dose of ultraviolet radiation than can be safely accepted for long periods.

- Singapore water is quite pure, but in many parts of the world, one must boil the water, and even filter it, before drinking it.

- Metro trains loudly warn while stopping at every station, *"Please mind the platform gap"* (between the edge of the platform and the carriage step as the passengers get off). How many hear them consciously anymore? Does it register in the mind?

- Children get their toes caught in escalators.

- People are hit by vehicles in pedestrian crossings.

- People get hit by "killer litter" dropped from higher floors.

Yet, we take most of the hazards in our stride, take our survival from them for granted. They become part of our life.

To manage hazards they must first be identified, simply because whatever hazard is not identified may develop into accidents, for which we will not be ready.

Author is excluding from any discussion suicides and the "voluntary risks" taken by many people such as smokers, rock-climbers, etc.

4.2. RISK

'Risk' is what happens when a hazard is realized, meaning when the potential danger becomes actual danger, affecting a person, an object, the environment, our reputation, or the bank balance.

Risk is the combination of many contributing factors, but two major factors stand out as the key ones:

1. The likelihood of occurrence of any mishap; and,
2. The severity of its consequence.

Specifically, risk is when your worker is likely to have an accident, or your property or environment is likely to be accidentally damaged, here and now. Not in the next block, not five years from now.

4.3. INDEPENDENCE OF LIKELIHOOD AND SEVERITY

We must always remember that likelihood and severity are independent factors contributing to risk. An event may be very severe but very unlikely; another may not at all be severe, but quite frequent.

We tend to think of something that makes big headlines as something that happens quite often. Aircraft crashes are a very common example.

Another example is deaths due to falls from scaffolds. They make news headlines and attract the highest penalties. But the likelihood of someone falling to death is really not very high.

Annually if 25 workers fall to death in about 5000 construction sites across the (Singapore) island, this works out to only one death in about 200 sites per year, that is, just one death in 200 years for any one site.

To reiterate, severity is the extent of consequence of a hazard if and when it occurs, regardless of the likelihood of its occurrence, but considering existing safeguards against severity.

Likelihood is the assessment of whether something will occur, and if so, how frequently, regardless of the severity of its consequence, but considering existing safeguards against likelihood.

4.4. SAME HAZARD, DIFFERENT OUTCOMES

Theoretically, our estimate of hazard, risk – almost anything – may range from 'Nothing' (or 'Never') to 'Everything' (or 'Always'), that is to say, from 0% to 100%. But in reality, the consequence or occurrence of any event or situation can neither be 'Nothing' or 'Never' (0%), nor be 'Everything' or 'Always' (100%).

So, if something seems like 'Nothing', we should say 'Very, very little' and estimate it at (say) 1-5%. If something seems like 'Everything' we should say 'Very, very much', and estimate it at (say) 95-99%.

The same hazard may result in different risks. Some examples:

(a) Same hazard, but different impacts, depending on user and use:

Gun is a hazard. The risk is very low (to a law-abiding citizen) when it is with a cop, but very high in the hands of a criminal.

(b) Same hazard, different severities, depending on race / genetics:

The sun is so essential to all life on earth. But it is also a hazard which may lead to heat prostration, sun stroke, skin cancer, and so on. Westerners are more prone to skin cancer due to solar radiation than Easterners, and their workforce has regulations to protect them. But as long as Asian workers are not directly exposed to hot sun for long times the sun is not a hazard to them.

(c) Same hazard, different impacts, depending on improved technology and proper management:

A hazard may start as a high risk, but with proper assessment and mitigation controls, it can be tamed to become a low or at least a medium risk.

For example, working at height was a high risk job in the 1930s when the Empire State Building was being built, with workers not even wearing helmets or anchored waist belts.

Today, it is at worst a medium risk with low likelihood, capable of being managed well with training and supervision.

(d) Same hazard, different reactions, depending on familiarity with and exposure to the risk:

How the Empire State Building was completed with only five fatalities and even many other hazardous structures were built with safety records as good or better than today's is an interesting analysis of how familiarity and exposure to danger hardens and prepares us to manage the danger.

In brief, if we grow up with danger, by example and trial and error, we learn to cope and survive. We intuitively learn to recognize and avoid the hazards or at worst, endure a decreased level of damage.

(e) Same hazard, different outcomes, depending on

time and usage:

A hazard may have very low risk level to start with, but may worsen.

Routine use of a machine, without proper maintenance, and deteriorating over a period of time, may insidiously and without warning, grow into a high risk task.

(f) Same hazard, different probabilities, depending on safeguards:

A tiger is undoubtedly a major hazard. Out in the open jungle, it is almost certain death to humans. But in a zoo it is not at all a risk to visitors.

Over this background, looking at the picture of a brave (?) photographer 'shooting' a tiger in the wild (Fig. 4.1, top), we may call it a 'very, very high risk', say 95-99% risk for the man.

If the tiger is behind a zoo's metal fence (Fig. 4.1, middle) the man's risk drops to 'very, very, low', say 1-5%.

Now, let us consider a doorway and gate into the zoo enclosure. (Fig. 4.1. bottom.) The caretaker may forget to latch or lock it; the latch or lock may be flimsy; the hinge may be rusted, and so on.

With so many potential problems, the risk of the tiger escaping and attacking the photographer (and other unfortunate visitors) is now 'medium', say between 1-5% and 95-99%, depending on the controls in place, the maintenance regime, and the disciplined behavior of the attendants – in Singapore, the author (having lived in other countries) would say 5-10% risk.

Fig. 4.1. - Tiger as Hazard and Risk

(g) Same hazard, different impacts, depending on age and experience of personnel.

A new worker in his enthusiasm and over-confidence, may overlook basic dangers from his lack of experience. An experienced older worker may be unable to carry out his duties with the care and strength needed for safety.

(h) Same hazard, different perceptions, depending

on job scope and context.

One death in a construction site may be unavoidable. In an office it would be unconscionable. On a battlefield it would be quite acceptable.

Hazards and risks are both subjective. Risk perception depends on the company (the word being used for any organization hereinafter), the product, the process, the people, the pressures, the context, the time, the environment, and familiarity with the risk. So hazards must be watched and analyzed with reference to circumstances and many other factors.

Hazards must also be reviewed and re-evaluated on a regular basis and whenever there is any change in a product, process, or personnel.

4.5. RISK ASSESSMENT AND MANAGEMENT

Risk assessment is a formalized procedure to determine:

1. What, if any, potential dangers exist in a particular segment of a project;
2. If and how they will develop into actual risks which may lead to accidents; and,
3. How bad the resulting accidents may be.

The first consideration constitutes "Hazard identification", involving some variation of job safety analysis to detect potential dangers, applied to stages of a project divided into distinct tasks or jobs, each job small enough to identify and describe the hazards in a finite number of related steps.

The 'if' in the second item refers to the evaluation of the probability of occurrence of the mishap translating into a risk. The 'how' in it refers to the consequence of the mishap to health harm, property or environmental damage, etc., when the mishap occurs.

The third item constitutes the actual risk assessment, based on the combination of hazard likelihood and severity estimates.

From this point on, actual risk management takes over, and involves risk control decisions, implementation, record-keeping, review, communication, follow-up etc.

A main criterion that should govern the choice of a hazard for further consideration of its translating into risk is that it be 'credible' – meaning that it be feasible and possible, even if it is not probable.

A corollary criterion is that the corresponding risk should be controllable, by the individual, company, or the authorities.

For instance, terrorist attack is a very credible hazard in most countries today. Singaporeans too should not brush it aside as improbable. At the same time, whether terrorism will affect the particular workplace and the specific project needs to be considered carefully, particularly because control at a company level may not be feasible or effective.

Likewise, nuclear bomb threat, while it may be real internationally, cannot be a concern for industrial risk assessment. Firstly, it is not a credible threat (because no nation can start a nuclear war and escape being destroyed itself – there are suicide bombers, but there cannot be suicide nations!) Secondly, there is nothing much a company can do if and when faced with it – apart from following government instructions.

However, emergency preparedness on the worksite, and inclusion of relevant controls in a major project in case of a bomb threat may not be such a bad idea.

Industrial espionage and internal sabotage, or the local equivalent of terrorism, may be a real threat in big and successful companies, and the risk assessment must include these hazards.

Non-credible hazards (for SMEs) are not discussed further in this book. But in reality, the consequence of any event or situation can neither be 'Nothing' or 'Never' (0%), nor be 'Everything' or 'Always' (100%).

Events which may happen a long time (say 3-5 years) after the scheduled finish date of the project, risks well beyond the capabilities of the company, etc. also should not occupy too much time of the RA team.

4.6. IMPORTANCE OF LESSONS LEARNT

The safety record of a company will depend upon how well it assesses and controls risks. Safety record is perfect only until the first accident.

A certain State in an Eastern developed nation made global news with its 13 accident-free months in 2005-2006. Its laudable record was unfortunately marred by five deaths in the 14th month.

This is not to say that the accident-free record was pure luck. But the message here is that no organization can rest on its laurels. We must keep on trying – aim at zero accidents, but keep monitoring and counter each and every indication of worsening of existing hazards or onset of new ones.

A good safety record would continue to help only if a cause and effect relationship could be found as to how exactly the accidents were reduced or eliminated.

Thus, after a long accident-free spell, it would be wise to examine all the parameters and compare them with the past records to isolate what, if anything, might be the reason for the improvement.

As example, UK attributes its record of no "serious" falsework collapse since 1972 to its adoption of the lateral sway resistance requirement of 2.5% of the vertical force, as against the 0.5%-1% previously.

Conversely, if there is a series of accidents, it would be even more worthwhile to find out root causes and extract common contributory factors from the data, so that we may avoid or control the risks.

Take the lessons learnt from past accidents and accident-free periods as a spur for greater effort.

You never lick the problem forever; you face it and control it all the time. You never arrive; you keep on running, for safety's sake, for your workers' sake, for the industry's sake.

–––

5. GET STARTED ... AND KEEP GOING

A risk assessor should not be like a lone ranger shooting from the hip. It actually takes a team to manage enterprise risk.

Nor should the RA team directly jump into the fray – it takes quite a bit of preparation. The process is an elaborate one that continues cyclically as long as a company is in the business – and possibly for three years thereafter!

Hence careful pre-planning is necessary for risk management becomes effective.

5.1. PRELIMINARIES

(a) Unit of risk assessment:

The unit of risk assessment is a 'job', comprising a fixed process with specific input, output, and trained personnel.

This is because not all parts of a project will be equally hazardous, and the different parts may have varying types and magnitudes of safeguards required.

Hence a project must first be divided into distinct phases or stages, and each phase must be divided into separate jobs.

The jobs must be well-defined and should not overlap

too much.

For example, in a building construction project:

- <u>Phases (or stages) of construction:</u> Excavation – base slab – basement construction – superstructure – ...

- <u>Jobs for basement construction:</u> Base preparation – falsework erection – formwork erection – concrete casting – ...

This is called 'scoping' of the project.

Finally the steps to execute each job should be listed for the risk analysis. Continuing the previous example:

- <u>Steps for falsework erection:</u> Set up sole plate and base plate – erect standards with cross braces – ...

(b) Starting up the process of risk management:

1. Form the risk assessment team for the phase – rather than for the project because different phases may require varying skills and experience. Get it approved.

2. Discuss the scope of the RM with management and other key people. Get a budget and time frame approved. Start modestly and build up, and not extravagantly or over-ambitiously, only to be forced to rush through, cut back or abort.

3. Start a "Risk Register", as in the following section.

5.2. RISK REGISTER

Start a "Risk Register", an actual physical file, plus a folder in your computer, to document everything your team will do as a risk assessor.

This register will be your reference point, anchor, and evidence of your noble intentions and hard work. This will be your professional proof and legal cover.

The authorities will ask for it, will want to look into it, and discuss details mostly on the basis of what is in it. If the risk register is as it should be and if your actions reflect its contents, it will serve as proof that your company has done everything possible to avert accidents.

Put another way, without a risk register, all your claims of sincerity and "in-principle" compliance will sound hollow. Without a complete and current risk register, it is an automatic violation of the Act.

For big projects, you may want to have more than one register, for distinct parts or stages of the project.

Computerization is fine. It will help you develop similar documentation for other jobs and projects, and it will be easy to make modifications and correct errors. But nothing like a hard copy to show off to officials!

The risk register should include the following and more:

- The composition of the Risk Assessment Team, endorsed by the management. Risk assessment is not a one man (or one woman) job, for reasons which will be explained later. Even inspectors will look for the RA team, not one risk assessor.

- Copies of, or at least references to, relevant designs, drawings, and contracts for the project or the stage concerned.

- Records of accidents and incidents ('near-misses') for the project or its stages. If your company has not had any such, borrow – not the accidents but the records – from a similar company or industry.

  The government can and does provide a lot of information. MOM, for instance, has in its website (*www.mom.gov.sg*) tabs for 'Statistics' and other leads, which should serve as a starting point.

- Site, time, and budget constraints

- A checklist (or 'prompt list') of safety-related items

- Brief but clear minutes of meetings

- Decisions made, and actions taken. Don't worry about errors or modifications. It is better to leave all of them in, as evidence of logical evolution and continuous improvement. No one can criticize mistakes and deficiencies as long as they are rectified.

- Copies of all the Risk Assessment Forms.

  Further guidance is provided in succeeding chapters.

5.3. RISK ASSESSMENT TEAM

How big a team? At least two, preferably three for a Small Enterprise, three to nine (but not much more than ten!) for a Medium Enterprise.

As the team will have a responsible and continuing activity, members must plan to stay on the job for at least a year or two. They must attend and participate in every meeting unless abnormal situations prevent it.

Don't pick your office mate or buddy. If the two of you think alike, you will miss even obvious dangers. Get someone with a different background, with different experience. For instance, if you are afraid of heights, pick someone who isn't, and vice versa. Pick someone from across the hall, or someone on a lower or higher floor, physically as well as metaphorically, that is to say, a junior or a senior to you, from a different division.

If your company is big enough, include representatives from all stakeholders, such as management representatives (NOT the CEO – he won't have the time!) designers, engineers, maintenance people, supervisors, foremen, contractor and sub-contractor representatives, suppliers, etc.

If possible, get at least one person from each specialty involved.

If and when necessary, get a consultant to advise – but only on a short-term and specific assignment basis, and not to run your company's RA and RM. Consultants won't normally carry the risk for your company.

It is best to develop in-house risk assessment capabilities anyway, so that you will have continuous and complete control of the risks.

The RA team must be approved by the management. One person is chosen as the team 'Leader', more as a spokesperson for the team rather than as the most knowledgeable or the most senior.

5.4. HAZARD IDENTIFICATION

Identify and record all hazards big and small, in the job selected for assessment.

This is the most important step in RA. A hazard not identified is a hazard not eliminated or controlled. Details of the identification process will be presented in a subsequent chapter.

A number of formal methods are available for hazard analysis: Job Hazard (or Job Safety) Analysis, Event Tree Method, Fault Tree Method, etc.

Many of these are used as hazard identification tools. But some of these methods are specialized and expensive in time and money, and thus considered beyond the scope of this book.

For SMEs, basic techniques such as listing the hazards for each job step or answering 'What If?' type of questions should be quite adequate.

5.5. RISK ASSESSMENT PROCEDURE

1. Evaluate the severity of consequence of the hazardous event or situation, if and when it happens.

2. Evaluate likelihood of occurrence of e hazardous event or situation.

3. Note all existing (or required) safeguards for the hazards chosen.

4. Combine the likelihood and severity to assess the risk level.

5. Group the risks under three or more categories: 'Acceptable', 'Unacceptable', 'Tolerable', and so on.

5.6. RISK CONTROL PROCEDURE

1. Shelve the risks considered 'Acceptable', for now.

2. Mark the risks considered 'Unacceptable' for further consideration.

3. Prioritize the risks rated 'Tolerable' in decreasing order of importance.

4. Establish controls for the prioritized risks according to a standard 'hierarchy', meaning, the order of decreasing effectiveness.

5. Assign specific staff and fixed dates for implementing controls.

6. Evaluate the residual risk after implementing additional controls, to determine the actual benefit from the additional controls.

This sequence of activities for risk assessment and management is depicted in the form of a flow chart in Fig. 5.1.

5.7. FOLLOW-UP PROCEDURE

1. Document each and every discussion, decision and action in the Risk Register.

Keep records for three years, to allow for review and evidence if and when an older case needs to be opened up in connection with a subsequent adverse consequence, or to establish a track record of compliance.

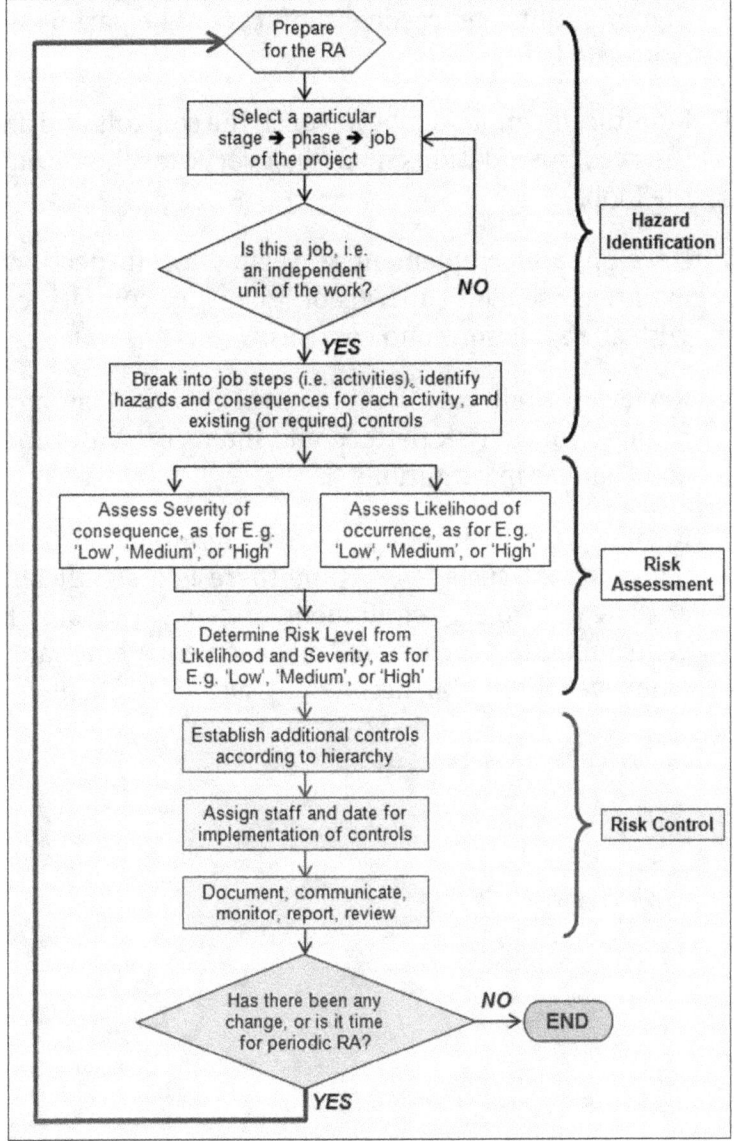

Fig. 5.1 - Flow Chart forRisk Management

2. Communicate all decisions to all the concerned individuals and groups, with appropriate explanations and supporting information.

3. Clarify doubts and resolve conflicts on the part of all concerned.

4. Monitor the implementation of all the controls and all the recommendations made by every stakeholder in the loop.

5. Develop and implement a regime of inspection, maintenance, and supervision to ensure compliance with all regulations and controls.

6. Review the effects of the controls, as well as any changes in the risk pattern and intensity, with time and changes in conditions.

7. Review and repeat the risk assessment and management cycle every time there is a significant change in project, equipment, process, personnel, environment, legislation, management, etc., and whenever there is an accident or serious incident.

———

6. PREPARATION

6.1. COLLECTION OF INFORMATION

The following information should be collected and kept available for ready reference and discussion during the risk assessment process:

- Site or plant layout plan
- Construction or process flowchart
- Details of work activities and/or trades
- Details of construction materials, chemicals, etc. used
- Size, shape, weight etc. of articles to be moved around
- Details of personnel involved in regard to safety requirements
- Details of equipment and tools used
- Relevant legislation, and codes of practice or specifications
- Relevant industry reports and media reports
- Inspection records
- Records of past incidents and accidents
- Health and safety audit reports
- Details of existing risk controls
- Feedback from staff, clients, suppliers or other stakeholders
- Safe Work Procedures (SWP)
- Other information, E.g. MSDS, manufacturers' manuals

- Copies of any relevant previous risk assessments

6.2. MAJOR FACTORS

(a) Scoping the project:

The team first defines the scope of the risk assessment to fit within the budget and time constraints. This must be approved by the management, to avoid problems later.

Then team members distribute the collected information among themselves and review the data, noting points that would have to be addressed during their RA.

They divide the project into jobs as already mentioned and distribute the job risk assessment responsibilities to the different members according to expertise and needs.

(b) Acceptability and Unacceptability:

Bounds of acceptability and unacceptability of hazards and risks must be tentatively laid down, even if subject to subsequent revision. These boundaries are dependent on the industry, the project, product, personnel, funds, and management policy.

The reason this is important is that acceptable risks are considered 'Low' and not demanding specific attention, while unacceptable risks are considered 'High' and to be avoided at all costs.

(c) Action points:

Personnel from various levels must be interviewed to evaluate their experiences and to document their comments.

Points must be freely brought up and frankly discussed, so that there will be no argument or back-tracking once the RA starts.

Based on all the preceding, a checklist must be developed. If one is not already available, MOM and/or the Internet may yield one.

If it is an ongoing project, a site visit is essential.

If it is a project at planning stage, a brainstorming session would firm up the hazard list.

(d) Let us not vote on human safety!

If the majority in a team votes a hazard as of 'Low' severity or likelihood, but the minority – even a lone maverick – holds out for 'Medium', it is not the majority opinion that should count. Where life or other great loss is involved, voting is not the way to go.

Even if unanimity is not possible, a discussion of pros and cons of alternatives, a consensus (overall agreement in principle) should be arrived at.

If the majority is not able to convince the minority about its stand (without undue pressure!) then the team should go with the worse of the assessments, 'Medium' in the example mentioned above.

This is not a permanent cost problem, because by the next round of risk assessment, enough experience would have been accumulated for one or the other side to change its stand.

6.3. DISTRIBUTION OF RISK

In a climate of accountability for risk by those who cause the risk, small and medium enterprises would do well to distribute the risk to the source agents of the risk.

Thus for instance, if the main contractor employs a sub-contractor to supply scaffolds, then the main contractor should specify, by contract, that the sub-contractor must carry out the RA for the work he is employed to do, namely to supply and erect the scaffold at site.

Likewise, the risks during dismantling and removal of the scaffold also should become the particular sub-contractor's responsibility.

Only after the sub-contractor hands over the erected and approved scaffold to the main contractor should the main contractor become responsible for the risks during use of the scaffold. His direct responsibility should also cease when he has completed the permanent work and hands over the scaffold back to the sub-contractor.

If the sub-contractor does not wish to or cannot conduct the RA, then the main contractor may accept the responsibility on the clear (contracted) stipulation that the costs of the RA and related accidents would be borne by the sub-contractor, and that the supervision and control of the risks would vest with the main contractor at all times.

The main contractor of course will be responsible for coordinating and integrating all the risks, even when specifics are handled by others.

Wherever possible, risk must be transferred to other

specialists who can handle them better, and/or covered by appropriate insurance.

As these involve legal matters, a lawyer should be consulted for actual contract terms. Author has only suggested principles and ideas.

6.4. RISK MANAGEMENT IN PRACTICE – 3

Do cell phones cause petrol-station fires?
– VERY, VERY UNLIKELY!

There is no scientific proof that cell phone use at petrol stations causes fires. It is true that static build-up due to any cause – including cell phones – can cause such fires.

But most static electricity has been traced to people in synthetic fiber suits (mostly women, in USA) swinging off their car seats while getting out, and then touching the fuel nozzle to the car body, creating a big enough spark to set off the fire.

Many petrol stations abroad have static discharge panels for this, as shown at the right part of Fig. 6.1.

Yet, the consequences of a petrol station fire are so catastrophic to so many people and so much property, that the following rule, called the 'Precautionary Principle', applies:

"If severity is extremely high but the control is simple, then impose the control – even if the likelihood is extremely low"

Switching off the cell phone for a few minutes is not a

big sacrifice, considering the one-in-a-billion chance of setting off a fire with it.

Fig. 6.1. - Petrol station fires due to cell phones or static electricity?

BUT, use of cell phones by workers at a site is still a NO-NO!

———

7. HAZARD IDENTIFICATION

7.1. PROPER MINDSET

The first thing all prospective risk assessors and managers must learn is to become and continue to <u>be</u> aware of each and every potential and real danger around us.

A risk assessor needs a special mindset to be able to identify hazards.

He must see more, hear more, smell more, taste more, and feel more than the average person; understand more than others about what is going on around him, look under the carpet and behind the screen, and read between the lines, so that he may track beyond the obvious.

Hazards may not announce themselves. For every hazard which is clear to even the untrained eye there may be another not so apparent.

To illustrate, I usually show a picture (left, Fig. 7.1) of a flight of steps at a children's park, and ask the audience what, if anything, is wrong with the scene.

Many participants see the big gap between the top bar and the steps, and comment that people may fall through it. That is all.

Then I show them the close-up of my hand trying to grasp the so-called handrail (right, Fig. 7.1). (I call it a 'hug-rail'!)

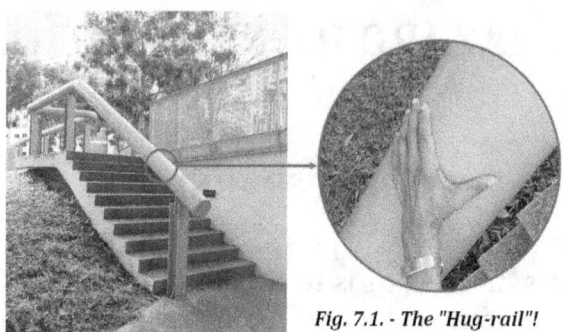

Fig. 7.1. - The "Hug-rail"!

This is what I mean when I say that the risk assessor must learn to look for the latent, the potential, the hidden, un-obvious hazard!

He must overcome any of his own known prejudices or biases, and phobias, so that he may know what can harm others.

He must expand his horizons beyond the work site and office hours, so that looking wider and deeper, and searching for potential dangers becomes a habit and a way of life, than a mere duty to earn a salary.

He must be concerned yet detached, firm but pragmatic in his quest.

He must be skeptical without seeming suspicious, questioning without appearing to dominate, and conservative without being pessimistic.

He will soon find that the new attitude brings him better understanding and more graceful acceptance of personal and home life, and of life itself.

In short, hazard identification is the most critical step in the chain of job safety analysis – a hazard not identified cannot be assessed or managed, and hence will remain a threat forever.

7.2. CREDIBLE THREATS

One warning may be appropriate here: The hazards we identify and analyze must satisfy the test of 'credibility', namely it must be in the realm of normal experience, and not imagined fantasy nightmares.

For instance, while Presidents and the military of powerful nations may worry about nuclear warfare, the industry professional need not include nuclear attack in his list of hazards. Likewise for an asteroid crashing into the earth, because there is nothing much a private organization can do about it.

A terrorist attack may not be a credible threat for a construction site but it can be one for a chemical plant.

At the other extreme, minor everyday happenings do not deserve a place in risk analysis,, especially if they are freak accidents which are not reproducible. Just because a lady tripped on gentle well-lit steps leading to the ladies room and died hitting her head on the landing does not mean that we should think of safety harnesses for all users!

7.3. IDENTIFICATION BY TYPES OF HAZARDS

Hazards may be identified by their type, or category:

- Physical – E.g. Falling from heights, slips and trips, fire

- Biological – E.g. Bacteria, fungi, viruses, dengue, SARS, H1N1

- Chemical – E.g. Acids, alkalis, solvents, toxic gases

- Electrical – E.g. Frayed wires, multiple outlet overload
- Mechanical – E.g. Damaged equipment, forklifts, cranes, unprotected machinery
- Environmental – E.g. Bad weather, excessive noise, smog, radiation
- Ergonomic – E.g. Manual handling, repetitive work, prolonged standing, awkward postures
- Psycho-social – E.g. Overwork, poor communication, bad supervision

7.4. IDENTIFICATION BY CONSEQUENCE OF HAZARDS

The most critical consequence of hazards, that is to say the most important result of accidents, is obviously their effects on humans, namely physical harm, such as sickness, injury, or death.

Beyond this, there could be many other consequences:

(a) Property damage:

E.g. Damage to cars in an automobile accident; loss of valuable paintings in a fire.

(b) Environmental damage:

E.g. Chemical plant toxic gas release into atmosphere; polluted effluent mixing with water supply.

(c) Time delay:

E.g. Failure of an important machine in a factory

mass-production line; stop-work order for a violation.

(d) Reputation damage:

E.g. News headlines of frequent high-profile accidents (Singapore, 2004), or frequent plane accidents (Indonesia, 2007); consequent lowering of safety statistics ranking in the industry and nation, with potential for cancellation of international contracts, reduction in tourist traffic, etc.

(e) Legal penalty:

E.g. Fines, suspension or dismissal, loss of certification, or jail terms, depending on violation.

(f) Financial loss:

E.g. All preceding items; workers' compensation payments; penalties or lost deposits due to missed deadlines.

Indirect and unknown costs can be many times the direct costs.

Sadly but realistically, every adverse consequence can be reduced to dollars and cents. This is understandable because money is the simplest and most effective yardstick to compare relative value of gain or loss.

Even human injury and death can be reduced to dollar equivalents:

A simple fracture may amount to a few thousand dollars. A worker's life is worth about $50,000. A professional's life value could go up to $1 to 2 million. Loss of an opera singer's voice may rate $10 million!

Time too may be converted to dollars: Cost of a 20-minute stalling of a MRT train during peak hour can be computed as: (Time in hour, 20 min./60) × (No. of passengers, say 1800) × (Average wage per hour, say $15) × (Factor to allow for indirect and hidden costs, say 5), that is $45,000 (assuming most of them work).

We have ignored loss in MRT productivity, major losses due to missed interviews or doctors' appointments, cancelled contracts, or other 'opportunity costs'.

Thus, a strong business case may be made for management to invest in safety, even purely on financial grounds, benefits out-running the costs.

7.5. SOURCES OF HAZARD INFORMATION

(a) Accident and incident statistics:

Seek them out; borrow statistics if your company doesn't have any. Refer to accident reports and statistics, in-house, or from similar industries, from published literature, or from MOM database.

"Those who cannot remember the past are condemned to repeat it." said George Santayana (1905).

(b) Annual reports:

Not just yours, but any and all similar industries in the country and region.

(c) News media:

TV, newspaper, and magazine reports of recent past. Again, not just local, but regional.

Remember however, accidents make front page headlines, mundane details and corrections are in small print in the last page! Look out for follow-up reports.

(d) Equipment and process manuals:

These often highlight the dangers in their commissioning and initial use, and how to avoid adverse consequences. E.g. Computer manuals give sound ergonomic advice on posture at, and use of computer equipment.

(e) Material Safety Data Sheets (MSDS):

While known to be important, these are often not studied or their recommendations followed by all concerned. They contain lists of sources of potential dangers in handling and using the material, consequences of their occurrence, and first aid for them.

Nowadays, equipment safety data sheets are also covered along with materials in the catch-all term 'SDS'.

(f) Codes of Practice and Regulations:

These too are an often neglected source of information. Almost every requirement in these was included to eliminate or mitigate (meaning reduce the impact of) some hazard, developed from investigation of previous accidents or predicted from theoretical or experimental research.

Hence, all an assessor needs to do is to check compliance with all the applicable clauses, and note any deviation from any regulation as a hazard.

(g) Checklists:

Don't have any checklist in-house? Try the MOM

website. Industry leaders are compiling exhaustive checklists.

Even if none is available, or if it is a novel or locally unfamiliar job, there is no need to panic. Open any Internet search engine, type your job name and 'checklist' (or other) keyword. You will get so many websites that you cannot read all of them in years.

For instance, typing "scaffold checklist" (without quotation marks) into Google gave (in Sept. 2016) more than a million sites.

Download the first few, collect common items, delete inapplicable items such as sunscreen ointment for Asian workers, cold weather precautions, etc. which are not applicable to your situation.

You have your own checklist!

(h) Recall:

From the experiences of the team members.

(i) Discussion:

With experienced field engineers, supervisors, foremen and workers, and seek their comments and suggestions, especially for infrequent activities.

They are the ones on the firing line, facing the dangers and the consequences of their tasks ... they are the ones who would know by direct experience.

(j) Brain-storming:

By the team, with consultants, etc..

(k) Interviews:

With resource persons, and (if necessary) experts.

(l) Walk through:

Of the entire site, for an ongoing project, or a similar site for a first-hand experience.

7.6. THE QUESTIONING APPROACH

The simplest and most common technique for securing information from sources listed in the previous section, is the questioning approach.

'What if?' is in fact a formal hazard analysis procedure. Excellent results may be achieved by questioning others as well as ourselves:

- Is there a source of harm here?

- Who will be harmed, how, how many?

- When, why, and how does the harm happen?

- What if one or more safeguards we take for granted do not work or are not practical? ...

- What if ...? What if ...?

7.7. JOB STEPS

Identification of hazards implies as a pre-requisite, the tabulation of the various separate activities or steps which make up the job, as described in Section 5.1.

Except for prevalent adverse conditions (such as haze), each job step is (usually) an activity, and must have an action word (verb) to define it.

For example, 'chemicals' is not a job step because it is not an activity; 'mixing chemicals' is. 'Ladder' is not a job step; 'climbing ladder' is. Hence, a job step is often referred to and tabulated as an 'activity'.

Each activity must have at least one hazard, and may have more than one hazard. E.g. Poor house-keeping on a work platform may cause (a) tripping of worker, and (b) damage to property.

Don't try to save rows in the RA form by combining hazards for a particular activity, because each hazard will generally end up with its own risk level, requiring its own control measure.

How many job steps? A simple rule is to note down as many steps as potential dangers you can identify. That is, if you cannot identify a hazard in a particular job step, there would be no point in recording it.

The author had the experience of a batch of trainees who listed ten hazardous activities to just reach and open the door of the classroom to go out – including the projector falling on the head, tripping while walking, and breaking a wrist turning the handle of the door!

On the other hand, do not pre-judge a hazard as trivial and omit it. Note even the smallest hazard, and decide on its importance later.

Each job is different.

Don't set any limits on the number of job steps at start. If during the process, the number of steps grows too large (say more than 20-25) then check if the job can be split into sub-tasks.

7.8. RECORDING CONSEQUENCES

Some RA forms include another column after 'Hazards', for 'Consequences' or 'Impacts'. Consequences usually define the severity of the hazard if and when it occurs.

Table 7.1 illustrates the sequence.

Table 7.1. Suggested columns for hazard identification

Activity	Hazard	Consequences
Climbing	Falling off	Injury, death

However, hazard and consequence may be combined into one entry, titled 'Hazard and Consequence', or even simply 'Hazard'. This single entry in the example case of Table 7.1 may read:

1. 'Falling off leading to injury or death'; or,
2. 'Injury or death due to falling off'.

When there are more than one consequence for a hazard, separation into two columns may help in clarifying details, assessing likelihood and severity levels, and in developing suitable controls.

It is good practice to enter them in different rows.

Some add a further column: *'Who might be harmed, and how'*, to help them assess severity better.

Thus, each company, each project, perhaps every job, may have its own hazard identification format.

The format itself may not matter as much as whether all potential dangers in that job have been identified.

7.9. MULTIPLE CONSEQUENCES

Just as one activity may have more than one hazard, one hazard may lead to more than one consequence.

The interaction between hazards and their consequences may be shown as in Fig. 7.2.

For instance the physical hazard of tripping and falling of worker carrying chemical bottle may result in:

1. Injury to worker,
2. Loss of bottle and chemical, and
3. Release of toxic chemical which may spread to surrounding areas, as pictured in the upper (solid arrow) set in the figure.

Fig. 7.2. - Hazards and Consequences

In an extreme case of an immigrant worker who has been stressed physically and emotionally (being away from his homeland and not being to communicate too well with his co-workers and superiors) may be so badly shaken by a small crisis like being criticized by his boss that his mind may not focus on his work.

While drilling into a wall, he may hit a gas pipe or electric line which he would have normally recognized, and end up with a chaotic mix of all possible consequences, as shown in the lower (dashed line) set in the figure.

It is critical that every hazard in the job must be identified along with its consequences, as otherwise, the corresponding risk cannot be eliminated or controlled.

7.10. A REAL-LIFE HAZARD IDENTIFICATION EXERCISE

In a recent two-day course of mine on RM, one of the participants – a middle-aged man – said this on the morning of the second day:

"Prof, I went home and thought about what you taught yesterday.

"Most evenings, my wife usually asks me to take a dish she has just prepared for supper, from the kitchen to the living room, where we would watch the TV serial together. – We have been married 25 years.

"Generally, I would pick the dish along the edge by my thumb and fingers of one hand, or hold the bowl between both my palms, and take it quickly to the living room.

"Most days, I would notice that the bowl was quite hot, but the living room was only about five metres away, and I managed quite well, even though it was often a relief to put it down on the coffee table, just before the heat got too uncomfortable.

"Yesterday, I used your 'What if?' trick, and asked myself, 'What if the bowl got too hot for me, and I dropped the dish?'"

Then the participant held out his right palm, and started counting off:

"*One:* It might have dropped on my feet, scalding my hands and feet (and who knows what else! – Author) – Physical injury

"*Two:* The dish so painstakingly prepared by my wife would have been wasted. – Time and money loss

"*Three:* There would be fuss and mess of cleaning up, and we would have missed out favourite serial. – Time and opportunity loss

"*Four:* I would never have lived it down, for years after. My wife would not have asked me to, or allowed me to, do anything else of that kind. I would have become the butt of ridicule among her friends (and enemies? – Author) – Loss of reputation

"*Five:* Last but not least, Prof, the dish is part of *her* family heirloom. Breaking it would have broken up the set. – Financial loss"

I could not help adding: "... *and it might have broken up your twenty-five years of marriage!* – Marital loss"

The participant finally ended the story with: "*From last night, I have begun to use the heavily padded thermally insulated gloves on both hands!*"

I truly believe that it was yet one more marriage I have saved!

[Dear reader: Why don't you look for such simple examples of everyday hazards, analyze how you handle them now, and then think about how you can improve on what you are doing? – Author]

7.11. RISK MANAGEMENT IN PRACTICE – 4

12 June 2006 OSH Alert
Occupational Safety and Health Alert

Fatal fall from scaffold

A plastering job ended up instead sealing the fate of a 37-year-old male Indian national construction worker last year. The worker fell from a height of about 5 metres from an open side of a scaffold while doing plastering work on the building façade of a terrace factory under construction. Subsequently, the victim succumbed to his injuries in the hospital.

Tip: Prior to work commencement, a risk assessment should be conducted in relation to the safety and health risks posed to any person carrying out the work.

––––––

8. EXISTING AND REQUIRED CONTROLS

Most of the time, we are not erecting a scaffold for a multi-storeyed building in the Amazon jungle. We are trying to reduce accidents in a sophisticated urban environment.

So there will already exist standard safeguards for every industrial activity.

The common practice therefore is to assess risk with the existing controls – after confirming that they are in compliance and in working order – and not in their absence.

This is easy to check in a project which is already ongoing – what the author refers to as the 'Inspection Mode' of risk assessment (RA).

In this situation, a safeguard is either present or not present. From direct observations, the likelihood and severity of the hazard can be assessed from lowest to highest, depending on what safeguards exist or do not exist.

Thus, if a working platform does not have a guard-rail, then the likelihood of falling is very high; if it has, the likelihood is very low.

Checking for and recording of available safeguards are best done while hazards are being identified. So this item may be included in 'Hazard Identification' stage, though many group it in 'Risk Assessment' phase.

8.1. NON-EXISTING CONTROLS

The problem arises with new projects, the RA for which the author calls the 'Planning Mode'. According to the WSH Act, everybody is required to conduct RA before a project starts. So, the argument goes, for a project still on paper, there are no existing controls.

In the conventional RA procedure, the assessor will enter 'None' for 'Existing Controls', resulting in all or most of the identified risks being determined to be 'High'. This practice is known as 'Gross RA'.

In such a case, a lot of effort and time will be used up in entering in the 'Additional Controls' column, impressive details of hand-rails, machine guards, chemical filters, fire extinguishers, helmets, etc. etc.

However, as soon as the contracts are let out, supplies start coming in, and the temporary works begin to go up, a quick RA exercise must be done to keep track of controls existing all of a sudden ... leading to moving all or most of the controls from the 'Additional' to the 'Existing' column, automatically transforming (most) former 'High' risks to 'Medium' or 'Low' risks. This then becomes 'Net RA'.

While there is nothing wrong with this procedure, author wishes to ask the RA team and the management: *"Can you start – will you be allowed to start – the project without the essential safeguards?"*

Obviously the answer will be 'No'. In other words, without the presence of mandatory safeguards and/or the commonly used, well-known and accepted controls according to a well-understood hierarchy, permits-to-work will not be issued.

So, no advantage will be gained in omitting the required safeguards during the initial RA exercise, self-correcting as such omission may be.

Author's recommendation (as is also the standard practice in many companies) is to expand the definition of the term 'Existing' control (in the sense of 'existing in Codes and Regulations') to include 'Required' or 'Mandatory' control as against 'Optional', 'Specialized' or other control.

In fact, it may be a good idea to re-write the heading of the 'Existing Controls' column as 'Existing/Required Controls' and then there will be no confusion!

Hence, the RA team should refer to all applicable Codes of Practice, regulations, standard SWPs etc. during their preparation stage, and include the essential safeguards under 'Existing (or Required) Controls' – and not let them be brought in as an after-thought!

No one should object to this, as it will save everybody a wasted round of RA. After all, the primary aim is to see that first all regulation-required safeguards are in place, and then additional safeguards are set to take care of exceptional or special hazards.

On the other hand, while listing 'Existing (or Required) Controls', it is not necessary to include any non-mandated (that is, exceptional or special) safeguard – unless the company has stipulated it also.

Showing exceptional and special safeguards under 'Additional Controls' has the additional value of attracting positive attention to them, from project staff and authorities alike.

8.2. WHAT EXISTING CONTROLS AFFECT

There is often confusion about whether controls reduce severity or likelihood.

Severity is defined as the extent of harm <u>if and when</u> the hazardous event occurs. This does not involve any question of likelihood.

However, it is psychologically natural to confuse between *"Will the worker fall?"* and *"How will the worker be hurt if and when he falls?"*

Typically, we look at guard-rails, toe-boards, etc. protecting a worker from falls, and immediately assume that the severity of a fall would be 'Low'. But the question should be: *"If and when the worker falls, how severe will be the harm?"* The answer now will be 'High'.

In case you wonder why we should worry about the severity of a worker falling when he is not likely to fall, the truth is that workers do climb up on mid-rails or even top-rails to access a point just beyond their reach from the platform.

Or sometimes the guard-rails may give way either due to poor design or workmanship, or due to an unexpected and excessive lateral force.

It is also true that most workplace safeguards reduce likelihood of harm rather than its severity.

So the effect of any control should be judged on a case by case basis for whether it influences the severity of consequence or likelihood of occurrence.

We will discuss this in detail when we discuss the controls.

8.3. SEQUENCE OF RECORDING

The reasons for listing various hazards for each activity and the different consequences for each hazard separately have been already explained.

Table 8.1 shows a typical layout of the Hazard Identification segment of a Risk Assessment form illustrating the sequence.

Table 8.1. Typical format of hazard identification portion of RA form

No.	Activity	Hazard	Consequence	Existing / Required Control	... and so on
1.	Activity 1	Hazard 1.1	Consequence 1.1.1	Control 1.1.1	...
		-,,-	Consequence 1.1.2	Control 1.1.2	...
		Hazard 1.2	Consequence 1.2.1	Control 1.2.1	...
2.	Activity 2	Hazard 2.1	Consequence 2.1.1	Control 2.1.1	...
			... and so on		

The exact order hazards are assessed and information is recorded may not matter. But author suggests that the entries be recorded job-step by job-step (row-wise), rather than all the activities first, then all corresponding hazards, next consequences, and so on (column-wise).

The simple reason is that in the column-by-column entry sequence, the assessor will have to re-think the information sequence all over again. However, in the row-by-row entry sequence he will be recording all the relevant information while the logic is still fresh in his mind.

Remember: The common habit of dashing off one line of analysis for each activity is very dangerous. Like a tree has many branches each with its own set of branches, one activity can have many hazards; one hazard may have many consequences.

You miss one sub-term in one item, and that will be the one to come back and bite your back when you are not looking!

———

9. ASSESSMENT OF LIKELIHOOD

The order in which we discuss likelihood and severity of a hazard does not matter, because the two factors are independent of each other, and we are going to combine the two together to determine the risk anyway.

Likelihood is the estimate of whether a hazard will become a risk here, now, and to your worker, and if so, how frequently or rarely. It is often referred to as 'Probability', and if tagged with time frames (as, for instance, so many fatalities per year) referred to as 'frequency'.

Although a minor point, the fact that both 'Likelihood' and 'Low' are commonly abbreviated as 'L' may cause some confusion while field staff fill out a RA form.

This can be a problem even when showing the expansion of the abbreviations in the form, which the author believes is a must.

To distinguish which 'L' of the two words he is referring to in his RA form – shown in his worked examples in Chapter 18 – the author had to mark 'Likelihood' as 'L*' and 'Low' as 'L'.

Author also suggests a simple change from the word 'Likelihood' to 'Probability', abbreviated to 'P', would avoid all this confusion.

For those who are already using 'Likelihood', author suggests that they take extra care in interpretation of the abbreviations.

9.1. RANGE OF LIKELIHOOD

Likelihood cannot be a 'Never' (0%) or an 'Always' (100%) for any event or condition.

If there were only two cars in an area of 120,000 km^2, will they crash into each other? No way, right? Yet, there were only two cars in the state of Pennsylvania in 1896, and they crashed into each other.

What is the probability that a perfect coin perfectly flipped an even number of times will fall exactly half heads and half tails? By all rights we should get it 50-50, but we rarely – note I don't say 'never' – do!

That is why we say the likelihood is 'Very, very low' when we want to say 'Never'; we say 'Very, very high' when we want to say 'Always'.

9.2. LEVELS OF LIKELIHOOD

Just classifying anything into two groups, say 'Low' and 'High', is also unrealistic.

Life is not just black and white, not just 'yes' and 'no'. There are many grays in between, and often a 'maybe' or a 'depends' that bridges 'yes' and 'no'.

The minimum number of levels should be three. It can be more.

For starters, the assessment of likelihood should group hazards into one of three levels.

Various names are used to indicate the different levels: 'Certain', 'Frequent', 'Occasional', 'Rare', etc.

'Highly infrequent' can be very confusing to an immigrant (or even a local) not sufficiently familiar with English. Whatever term is used, it must be carefully and unambiguously defined, with examples.

As already explained, we will use the simplest of words ('Low', 'Medium', 'High') to describe levels of likelihood.

For instance, if routine tasks can be broadly divided into three levels of likelihood, but there are occasional events that happen much less frequently than the lowest level or much more frequently than the highest level.

Then one may create an extra level to catch the likelihood of this event, and call it 'Very Low' or 'Very High' as the case may be.

Thus, if once a year, once a month, and once a day were 'Low', 'Medium', and High', then something that happened once during the lifetime (of ten years or more) of a piece of equipment even with significant consequences can be designated 'Very Low' likelihood, for further evaluation. Constant noise may be rated as of 'Very High' likelihood.

In most of what follows, we will be considering only three levels. But, for more than three levels, comments for 'Low' and 'High' levels will apply respectively to the lowest level and the highest level in the set.

After gaining experience in assigning hazards to different levels, a team can venture to suggest more than three likelihood levels. Additionally, planning for more than three levels should be based on need, capability, and resources to define and implement the extra levels.

9.3. ASSIGNMENT OF LIKELIHOOD LEVEL

How does one assign a likelihood level to a hazard?

A simple technique is to note that 'Low' likelihood means the probability or frequency of the occurrence is 'Acceptable' under normal circumstances, and that 'High' likelihood means the probability or frequency of the occurrence is 'Unacceptable' under normal circumstances.

First, identify all hazards in the particular job.

Group as of 'Low' likelihood whatever occurs very rarely or has not happened in recent memory – at the same time, not impossible that it could happen. A power outage is a very unlikely event in Singapore.

Group them as of 'High' likelihood when something happens all the time or so frequently that one must be always prepared to face it. Some 8 out of 10 visitors to the author's apartment hit the stairs' last step which is about 5 cm higher than the rest – embarrassingly frequent!

Anything between these two extremes is 'Medium' likelihood. It may range from events or situations that occur occasionally to frequently.

In the case of three-level classification, defining the two extremes automatically defines the 'Medium' as consisting of those items not in the first two lists.

For likelihood at a construction site, deaths are 'Low', and minor cuts and bruises are 'High'. So, cuts requiring stitches, minor amputations and fractures would be of 'Medium' likelihood.

Table 9.1 gives typical definitions for the three levels for likelihood, sourced from Singapore regulations.

Table 9.1. Specimen description of likelihood levels

Likelihood	Description
Remote* ('Low')	Not likely to occur; not occurred in anyone's memory. E.g. Once in 10 years.
Occasional* ('Medium')	Possible, known to occur. E.g. Once or twice every year.
Frequent* ('High')	Common or repeated occurrence, all the time, can be expected to happen. E.g. Once every week.
* = MOM terminology. Bracketed terms are author's recommendation.	

The descriptions are general and not specific guidelines, because different industries and jobs will need different descriptions.

Author advocates specifying likelihood measures in terms of frequency (E.g. how many times a year does the mishap happen) or odds (E.g. how much percentage of the times the activity was attempted does the mishap happen).

The limits of acceptability and unacceptability of the likelihood of the hazard are also dependent on the industry, project, and other factors.

Thus, one death per year would be 'High' likelihood for a construction site, but 'Very Low' in a war and 'Very High' in an office.

Workers at a construction site may get so used to piling or jack-hammer noises that they may not even be conscious of them after some time. But the RA team must still indicate the likelihood as 'High', so that the workers may be protected from the consequences.

9.4. EXAMPLE OF MAN CROSSING DRAIN – LIKELIHOOD

Consider a person (assumed a normal adult, not a child or very short adult) pondering whether to cross a drain or not.

The likelihood of occurrence of a fall into the drain will depend on the width of the drain. Figure 9.1 depicts the man at a drain with different widths.

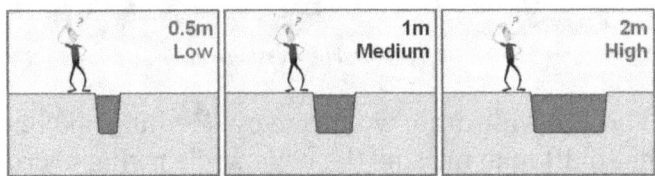

Fig. 9.1. - Likelihood of stepping or falling into drain

If the drain width *w* is:

500 mm : He can simply step across, with <u>low likelihood</u> of stepping into the drain.

1 m : He must take a deep breath and take a long stride across it, with <u>medium likelihood</u> of missing the other edge.

2 m : He must start well behind the drain, take a running leap and jump across, with <u>high likelihood</u> of falling into it.

A few months after the author had developed this example, exactly the same problem was faced – and wrongly decided – by a 76-year old man, who fell to his death crossing a 1 m wide and 2 m deep drain, as depicted in Fig. 9.2.

RETIREE, 76, FALLS WHILE TRYING TO CROSS THIS DRAIN

The New Paper
14 February 2007

Shortcut to death

He had used 1m-wide drain as shortcut for past 10 years to get home. Family says...

Photo: JOEL CHAN

Fig. 9.2. - Death while crossing a drain

The 1 m wide drain would have a 'Medium' likelihood of the (old!) man missing the edge while trying to cross.

9.5. DIFFICULTY TO DECIDE THE LIKELIHOOD LEVEL

Likelihood level is often quite difficult to assess. It is based on probability of occurrences, being matters of chance, more a guessing game than a scientifically predictable value.

Likelihood assessment depends heavily on historical records and statistics, and significantly on judgment, even on opinion. There is a lot of subjectivity and an element of chance involved in this assessment.

Historical records, statistics released by MOM, public

domain sources (like the Internet) etc., need to be consulted to enable the sorting of job hazards into the three or more levels.

Past statistics can be a good guide, but in many instances the data is insufficient or not robust enough to draw strong conclusions from.

Weather records have been collected for more than a hundred years in certain parts of the world, but even then one cannot be quite certain whether a weather prediction will apply to a particular place at a particular time or not!

Statistics for various activities abound in the literature and on the Internet. Most countries publish annual statistics of injuries and deaths from various causes, even separately for workplaces.

For instance, the site:
http://www.nsc.org/learn/safety-knowledge/Pages/injury-facts-chart.aspx
from the American National Safety Council gives the information listed in Table 9.2 on the chances of dying due to various causes in 2013. The odds are based on the odds of dying to any cause being 1 in 1. (Everybody has to go some time!)

It shows for example that (in 2013) that dying from a fall was only (133/113) or only about 18% more than dying from a motor vehicle crash (not restricted to work-related).

Taking another example, in a site with 900 workers, chances of deaths by falls (out of those exposed to falls?) would be (900/133) or about 7, which would be quite high – if they did not take proper precautions.

Table 9.2. Odds of Dying from National Safety Council, USA (2013 data)

Cause of Death	Odds of Dying
Heart Disease and Cancer	1 in 7
Chronic Lower Respiratory Disease	1 in 27
Intentional Self-harm	1 in 97
Unintentional Poisoning by and Exposure to Noxious Substances	1 in 103
Motor Vehicle Crash	1 in 113
Fall	1 in 133
Assault by Firearm	1 in 358
Pedestrian Incident	1 in 672
Motorcycle Rider Incident	1 in 948
Unintentional Drowning and Submersion	1 in 1,183
Exposure to Fire, Flames or Smoke	1 in 1,454
Choking from Inhalation and Ingestion of Food	1 in 3,408
Pedacyclist Incident	1 in 4,337
Firearms Discharge	1 in 7,944
Air and Space Transport Incidents	1 in 9,737
Exposure to Excessive Natural Heat	1 in 10,784
Exposure to Electric Current, Radiation, Temperature and Pressure	1 in 14,695
Contact with Sharp Objects	1 in 30,860
Cataclysmic Storm	1 in 63,679
Contact with Hornets, Wasps and Bees	1 in 64,706
Contact with Heat and Hot Substances	1 in 69,169
Legal Execution	1 in 111,439
Being Bitten or Struck by a Dog	1 in 114,622
Lightning Strike	1 in 174,426

However, we must be very careful in borrowing statistics from another country or another industry, and even more so, in using them. For instance, suicides ('Self-harm') amount to about 1 in a 100, which may not be so high in other countries.

Fortunately, right now we are dealing only with qualitative risk assessment, for which purpose, it is sufficient that we group them into relative levels of 'Low', 'Medium' etc.

There may be situations when nothing much is known about the likelihood of a certain event, or available data is contradictory or confusing. In such cases, it is better to start with 'Medium' level – corresponding to 50-50 chance of heads or tails in tossing a coin – and then let experience be the guide in subsequent assessments.

9.6. KICK-STARTING LIKELIHOOD ASSESSMENT

Fresh graduates, newcomers to the risk assessment field, or even veterans faced with a new project usually have trouble assessing the likelihood of a mishap. Author has found the following approach to work quite well and give the beginners confidence in their task.

1. List all occurrences of known (credible) mishaps for all activities in the job, in terms of so many times a year or other fixed period, or alternatively so much percentage odds of happening.

   Do not restrict your items to your organization, but include industry-wide happenings if you have common factors. Be sure to include unusual and infrequent activities also.

2. Define the criteria for m levels of likelihood, meaning how many groups you would like to classify all the mishaps in terms of their likelihood, and how you would like to group them.

   Example for 3 levels by frequency:
   (i)   Low – Once during the project or once in more than 5 years,
   (ii) Medium – Once a month to once in 5 years,
   (iii) High – Most of the time to once in less than a month.
   Example for 3 levels by percent odds:

(i)   Low – Bottom 15 to 20%,

(ii) Medium – Middle 60-70%,

(iii) High – Top 15 to 20%.

These are not formulas, and the boundaries have to be agreed upon within the RA team and by the management.

3. Arrange the hazards in increasing order of likelihood of their occurrences from Step 1, either in terms of the frequency or odds of occurrence.

We need not be too rigid in setting the criteria.

4. Split the arranged list into the number of levels chosen, generally with fewer in the lower and higher levels and more in the middle levels.

We should not split two mishaps with close likelihoods. There should be discernible breaks at the boundaries.

———

10. ASSESSMENT OF SEVERITY

Severity is the estimate of how a hazard will affect the victims (or other valuable entities such as property, environment, reputation, etc.), if and when the hazard became a reality. It is often referred to as 'Impact'.

Severity is much more predictable than likelihood. It can be measured by tangible, generally quantifiable parameters – like so many days of medical leave, or so many dollars of medical expense.

10.1. RANGE OF SEVERITY

As already mentioned, in real life, there cannot be a 'Nothing' (0%) or an 'Everything' (100%) for any consequence of any event or condition.

One cannot swear that no harm will come to someone even while doing something not obviously hazardous. A woman fell down a few carpeted steps in an Opera House and died (*"Woman tumbles down stairs on way to toilet during Esplanade show"*, TODAY (Singapore), 4 Jan. 2007).

One cannot also argue that an obviously very dangerous task will surely mean his or her death. A man fell 16 storeys and survived (*"Man plunges 16 floors and lives"*, The Straits Times, Singapore, 23 Jan. 2007).

So, we say 'Very, very low' when we want to say 'Nothing', and 'Very, very high' when we want to say 'Everything'.

10.2. LEVELS OF SEVERITY

As with likelihood, for starters assessment of severity should group hazards into one of three levels. Various names are in use to indicate the different levels, such as 'Minor', 'Moderate', 'Catastrophic', etc. Whatever term is used, it must be carefully and unambiguously defined, with examples.

The author reiterates his preference for (and advice of!) the use of simple words ('Low', 'Medium', 'High') to describe levels of severity.

If routine tasks can be broadly divided into three levels, but there are occasional events that result in much less harm than the lowest level or in much more harm than the highest level, then one may create an extra level for this event, and call it 'Very Low' or 'Very High' as the case may be.

If for instance, a bruise, a fracture, and one death have been termed 'Low', 'Medium' and 'High' severity, then for a job involving multiple deaths as worst case scenario, a new level 'Very High' may be created.

As with likelihood, we will be considering only three levels. But, for more than three levels, comments for 'Low' and 'High' levels will apply respectively to the lowest level and the highest level in the set.

As with likelihood, a team may choose to use more than three levels of severity, after gaining experience in assigning hazards to different levels and subject to the need, capability, and resources available, to define and implement the extra levels.

10.3. ASSIGNMENT OF SEVERITY LEVEL

How does one assign a severity level to a hazard?

As with likelihood, a simple technique is to note that 'Low' severity means consequence of the occurrence is 'Acceptable' under normal circumstances, and that 'High' severity means consequence of the occurrence is 'Unacceptable' under normal circumstances.

First, identify all hazard consequences in the particular job. At a construction site, injuries can range from a scratch to death.

Group as of 'Low' severity whatever can be accepted as reasonable part of the job, and unavoidable during the normal course of events; they should not cause significant worry or expense, and the situation should return to normal soon. E.g. A scratch at a construction site.

Group the hazards as of 'High' severity when they cannot be accepted as reasonable part of the job under normal circumstances, and are very traumatic to endure and expensive to rectify; they cause irreparable harm or serious long-term problems. E.g. Death at a construction site.

Anything between these two extremes is 'Medium' severity. It would cover many injuries that need immediate or fixed-term medical attention, but with eventual return to reasonable functionality within a reasonable time. They are unavoidable, but they cannot be ignored.

In a construction site, cuts that need stitches, fractures and minor (finger/toe) amputations etc. do happen almost on a regular basis. The resulting risks are considered 'Tolerable', to be carefully managed.

Actually, one does not have to define the 'Medium' level, in the case of three-level classification. Defining the two extremes 'Low' and 'High' automatically defines the 'Medium' as those not in the first two lists.

Table 10.1 gives a typical definition of the three levels for injury.

Table 10.1. Specimen description of severity levels

Severity	Description
Minor* ('Low')	No injury, injury or ill-health requiring first aid treatment only (E.g. minor cuts and bruises, irritation, temporary discomfort)
Moderate* ('Medium')	Injury requiring medical treatment or ill-health leading to disability (E.g. lacerations, burns, sprains, minor fractures, dermatitis, deafness, work-related upper limb disorders)
Major* ('High')	Death, serious injury, or life-threatening occupational disease (E.g. amputations, major fractures, multiple injuries, paralysis, occupational cancer, acute poisoning and fatal diseases)
*= MOM terminology. Bracketed terms are author's recommendation.	

In Table 10.1, the descriptions are general and not specific guidelines, because different industries and jobs will need different descriptions. They are more oriented to a medically oriented evaluation rather than a lay person's judgment.

Severity assessment based on this table presumes that the RA team knows from experience or records what kind of injuries had occurred during the various tasks.

To avoid confusion and misjudgment of the severity, author advocates the use of some authoritative quantitative measure of injury severity such as the duration of the victim's medical leave or absence from work due to the injury.

The actual boundaries of acceptability and unacceptability of the severity of the hazard will depend on the industry, project, and other factors.

10.4. HOW TO HANDLE AN ACTUAL INJURY AT SITE

Although not related to RA, what to do during an actual injury at site may be discussed here.

You cannot just stand there wringing your hands.

On the other hand, don't try to be a hero and try to 'save' the victim unless you are a paramedic! Apart from legal complications, the victim may get worse or even die by your intervention.

When it comes to severity of injury, engineers and in fact all personnel not specifically trained for medical emergencies should refrain from assessing or treating the injury.

Don't try to evaluate an injury on the basis of external bleeding or the victim's reactions.

Lack of bleeding may hide a serious internal injury. On the other hand, screams may not necessarily indicate serious trauma. Many may be too frightened, and some may even over-react (– to put it politely!)

When in doubt, it is better to assume that the injury may be serious.

Except for trivial ('band-aid') cases, call an ambulance, or send the victim to the site clinic if there is one, or to the nearest hospital. Let the doctor/clinic decide.

Needless to say, any and all first-aid available and possible should be rendered to the injured victim, including Cardio-Pulmonary Resuscitation (CPR) and Automated External Defibrillator (AED) if experienced persons and essential equipment are at hand.

10.5. EXAMPLE OF MAN CROSSING DRAIN – SEVERITY

Severity of consequence of a fall into the drain will depend on the depth of the drain. Figure 10.1 depicts a man at a drain with different depths. (Again, we are only considering normal adults.)

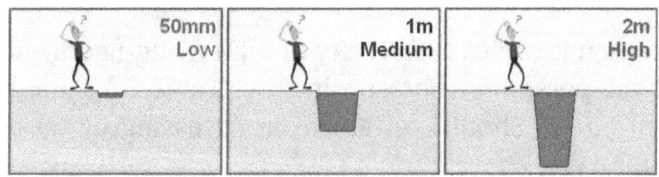

Fig. 10.1. - Severity of stepping or falling into drain

If he steps or falls into the drain of depth d while trying to cross it:

d = 50 mm : Water will just wet his shoes, <u>low severity</u>.

$d = 1$ m :   Water will reach to his waist level, messy though not life threatening, <u>medium severity</u>.

$d = 2$ m :   Water will be over his head, <u>high severity</u>.

In the actual case discussed in the previous chapter (Fig. 9.2) the old man crossing the 2 m drain could have anticipated the severity to be 'High'.

10.6. CONSEQUENCES APART FROM PERSONAL INJURY

Human injury and illness are the primary concern of every employer, and most organizations and authorities (such as Singapore's Ministry of Manpower) rightly focus on human health in assessing severity.

At the same time there would be other consequences of hazards, as has been discussed in Section 7.3, which deserve serious consideration.

Similar assessment techniques will hold for these as discussed for injury. Table 10.2 presents some typical severity levels for different consequences, with health and injury effects also included. (Legal requirements are not included, as they must be satisfied in any case.)

10.7. KICK-STARTING SEVERITY ASSESSMENT

As with likelihood assessment, fresh graduates, newcomers to the risk assessment field, or even veterans faced with a new project often have trouble assessing the severity of a mishap, physical or otherwise.

Table 10.2. Typical severity levels for different consequences

Conse-quence↓	Low Severity	Medium Severity	High Severity
Health and injury	First aid only, 1/2 to 3 days MC	> 3 days to < 6 months MC	6 months or more MC Death
Property damage	Inexpensive, and easily replaceable	Fairly expensive, and/or fairly difficult to replace	Expensive, or irreplaceable
Environmental harm	Affects a few people, temporarily, at a low level	Affects a few people seriously, or many people at a relatively low level	Affects many people, permanently or for a long time, at a high level
Time delay	A few minutes to a few hours	Many hours to a few months	Many months to a few years
Reputation damage	Inside page news, no court case, explanations accepted	Some headlines, short-term publicity, civil case, settlement out of court, damage control successful	Wide and blaring media cover-age, criminal case, resulting in loss of large contracts
Financial loss	Less than 5% of investment	Between 5% and 50% of investment	More than 50% of investment

Author repeats the same procedure as for likelihood assessment for severity here.

1. List all severities of known (credible) mishaps for all activities in the job, in terms of some logical measure of the consequence.

    This would depend on what type of consequence is being assessed. Injury may be assessed in terms of medical leave, property damage in terms of replacement cost, time delay in terms of number of days (or even corresponding loss of income).

Do not restrict your items to your organization, but include industry-wide happenings if you have common factors. Be sure to include unusual and infrequent activities also.

2. Define the criteria for n levels of severity, meaning how many groups you would like to classify all the mishaps in terms of their severity, and how you would like to group them.

    The criteria will naturally be different for different types of consequences. For physical harm the criteria may be based on duration of MC, for property damage it will be based on replacement cost, and so on.

    Table 10.2 gives examples for different types of consequences for a three-level assessment of severity.

    These are not formulas, and the boundaries have to be agreed upon within the RA team and by the management.

    We need not be too rigid in setting the criteria.

    Do not split two mishaps with close severities. There should be discernible breaks at the boundaries.

3. Arrange the consequences in increasing order of severity from Step 1, either in terms of the appropriate measure used to assess the severity.

4. Split the arranged list into the number of levels chosen, generally with fewer in the lower and higher levels and more in the middle levels.

10.8. RISK MANAGEMENT IN PRACTICE – 5

There are many risks we take – and we let others take – without knowing the consequences. One of them is manual lifting.

Many do not know that when we bend and lift any weight *w*, our backbone is subjected to a force up to 12 to 15 times of the total of that weight *w*, plus about half our body weight at half the distance of the load, i.e. about a quarter of our body weight.

This can be easily demonstrated by the lever principle, namely the lifted weight(s) times the distance(s) of the lifted load(s) must equal the force on the backbone times the diameter of the backbone. (*See* Fig. 10.2.)

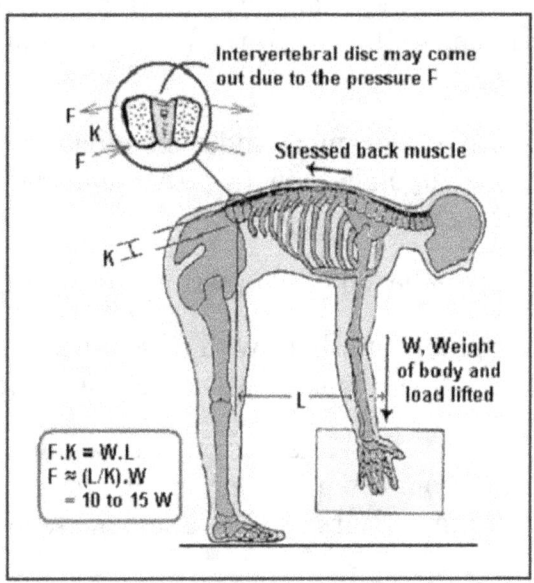

Fig. 10.2. - Bending and lifting a load

Nature has designed our backs to resist a certain force during our normal tasks safely.

To illustrate, consider a worker weighing 60 kg and having a backbone strength of 600 kg. (1 kg = 2.2 lb.)

From the 600 kg, take away (say) 15 times one-quarter body weight of 60 kg, i.e. 15 times 15 or 225 kg. That leaves about 375 kg. This divided by 15, i.e. 25 kg is what we can bend down and pick up safely on a repeated basis.

This is also the basis for limiting the weight to be lifted by workers (or anybody) on a routine basis, to 25 kg.

Do you know what your workers carry day in day out, week after week? Now that you know their limit, what can you do about it? What will you do about it?

Refer to my article: *"Risk Assessment of Manual Lifting"* in the 'Structural Engineer' Journal of the Institution of Engineers (Singapore) of Sept. 2006, p. 20-23, available from my website *www.profkrishna.com*, by clicking the 'Publications' thumbnail on its home page.

———

11. ASSESSMENT OF RISK

11.1. INDEPENDENCE OF LIKELIHOOD AND SEVERITY

As mentioned earlier, the two factors chosen as contributing to risk, namely likelihood and severity, are independent of each other. An event may be very severe but very unlikely; another may not at all be severe, but quite frequent.

We tend to think of something that makes big headlines as something that happens quite often. Aircraft crashes are a very common example.

Another example is deaths due to falls from scaffolds. They make news headlines and attract the highest penalties. But the likelihood of someone falling to death is really not very high.

Annually if 25 workers fall to death in about 5000 construction sites across the (Singapore) island, this works out to only one death in about 200 sites per year, that is one death in 200 years for any one site.

To reiterate, severity is the extent of consequence of a hazard if and when it occurs, regardless of the likelihood of its occurrence, but considering existing safeguards against severity.

Likelihood is the assessment of whether something will occur, and if so, how frequently, regardless of the severity of its consequence, but considering existing safeguards against likelihood.

11.2. COMBINATION OF LIKELIHOOD AND SEVERITY INTO RISK

Figure 11.1 depicts our man at the drain, now facing all nine possible combinations of three width (likelihood) options and three depth (severity) options.

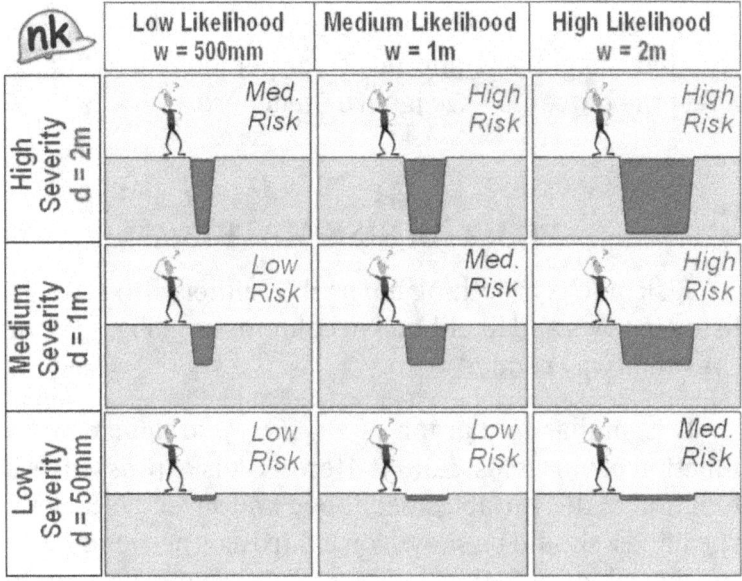

Fig. 11.1. - Man planning to cross a drain (Same as on Cover)

[The helmet at top left corner of the figure is my logo. – NK]

Note that the middle row of Fig. 11.1 is the same as Fig. 9.1 and the middle column of Fig. 11.1 is the same as Fig. 10.1 laid out horizontally.

The nine risk combinations of likelihood and severity, each of three levels range from extremely low risk (of low likelihood and low severity) to extremely high risk (of high likelihood and high severity).

The man's decision will – or should – be based on:

(a) Likelihood of his stepping into it while crossing [Sec. 9.5]

(b) Severity of consequence if and when he steps into it [Sec. 10.4]

In real life, other considerations like time of day, weather, height, age and health of person, etc. will also affect the outcome – we ignore them here.

11.3. THE RISK MATRIX

Risk matrix is nothing but a table, either with severity listed in rows and likelihood in columns as in Fig. 11.1, or the other way round.

It is similar to a graph of $z = f(x, y)$, to depict z as a function of variables x and y. Here, risk is expressed as a function of the variables likelihood and severity. Strictly speaking z should be shown on a third axis perpendicular to the plane of the x-y axes. However, in simple situations, the value or significance of every combination may be shown at the (x, y) location itself by words or numbers, as shown in Fig. 11.2.

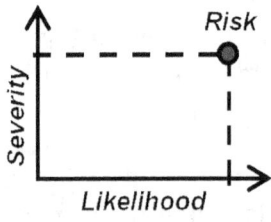

Fig. 11.2. Risk= f(L, S)

Although the problem is treated qualitatively and not quantitatively here, the principles are the same. Instead of the variables changing continuously, we consider them in finite steps, 'Low', 'Medium', etc.

Figure 11.3 shows the risk matrix for combinations of likelihood levels and severity levels. The region of risk combinations (of $m.n$ cells for 'm' likelihood levels and 'n' severity levels) is the "risk zone", within which the status of any risk combination of likelihood and severity can be indicated.

Severity↓	Likelihood		
	Lowest	→	Highest
Highest			MOST
↑		Intermediate	
Lowest	LEAST		

Fig. 11.3.- m by n risk matrix, 3 regions

Obviously, the combination of lowest likelihood (very infrequent) and lowest severity (minor harm) will be considered the 'Least' risk.

Likewise, the combination of highest likelihood (very frequent) and highest severity (major harm) will constitute the 'Most' risk.

The region between these two extremes is obviously 'Intermediate' risk.

In theory one can start with the lowest level for likelihood and severity at either end of a row or column, and proceed towards the other end in the appropriate direction, to reach the respective highest level.

Then, depending on the particular choice of lowest ends, the least risky cell will be at one of the four corners of the risk zone, as at the bottom left in the case illustrated in Fig. 11.3. Naturally, the most risky cell will be at the diagonally opposite corner from the least risky cell, namely at the top right in Fig. 11.3.

Risk categorization consists in dividing the risk zone appropriately into a certain number of sub-zones or 'categories', ranging from least risky to most risky.

The two corner zones, consisting of the least and most risky corner cells and possibly a few adjacent cells, are quite critical because they will be treated specially.

They will define the intermediate risk zone between them, which will be the zone we must manage. The intermediate zone may be sub-divided into further bands of categories.

Where there is no ambiguity, author will use 'Low' or 'Least', 'Medium' or 'Intermediate', and 'High' or 'Most', interchangeably.

11.4. RECOMMENDED CONVENTIONS

If each assessor and each company followed their own conventions, comparisons and cross-evaluations would involve considerable time and effort to arrive at common conclusions.

In fact, if an evaluator is careless or hasty and goes by cursory visual impression from the matrices, drastic errors may be introduced.

To avoid these traps, the author strongly recommends:

1. Companies should stick to the convention of left and bottom ends of the risk zone as the 'Low' likelihood and 'Low' severity cells.

2. When evaluating or comparing other risk matrices, assessors should double check their conventions, and mark the least risky cell (say with a single asterisk) and the most risky cell (with a double asterisk), and note this convention at bottom of the sheet.

3. The simplest wording consistent with clarity must be used.

4. All abbreviations and notations must be explained on every sheet the matrix is displayed.

If such care seems excessive, remember that a mistake may cost lives!

11.5. RESPONSIBILITY FOR RISK CATEGORIZATION

Assessment of likelihood, severity, and the resulting risk is mainly the responsibility of the RA team and any experts retained for specific purposes.

However, where decisions involve (a) funds available, and (b) company policy, as in definitions of likelihood and severity levels and risk categories, management input and endorsement will be required.

Whatever the subsequent contingencies and pressures, the categorization should not be changed during the course of project except under very extenuating circumstances.

After every decision on likelihood and severity levels for any hazard, the combined risk level is an automatic choice from the risk matrix already chosen and shall not involve any further decision-making.

11.6. NUMBER OF RISK CATEGORIES

The number of risk categories should be based on the number of distinct controls that can or must be developed to take care of the risks.

In no case should it be less than three, namely 'Low', 'Medium', and 'High', for the same reasons as stated for likelihood and severity.

Additionally, author recommends that it should be not less than the smaller of the number of levels of likelihood and severity, as otherwise the effort spent in sorting out the basic levels would be partly wasted.

One company used five levels of severity and five levels of likelihood, but ended up with only two categories of risk: 'Significant' and 'Insignificant'.

And because of the way they gave numerical values to various contributor consequences of hazards and added them to evaluate the risk, they ended up with certain risks involving death being tagged as 'Insignificant' – resulting in fatalities! This violates both the recommendations given in the previous paragraph.

In keeping with three-level likelihood and severity groupings, the most basic and common risk categorization is three-fold. For simplicity and clarity, these three categories also may be designated as 'Low', 'Medium', and 'High'. [*See* Fig. 11.4.]

This arrangement will be quite adequate most of the time, for SMEs, and also corresponds to the suggestion (– not requirement!) in Singapore MOM guidelines.

From Fig. 11.4, the drain 1 m wide ('Medium'

likelihood) and 2 m deep ('High' severity), denotes the intersection of the middle column and top row, which is rated as 'High' risk. Our man should surely <u>not</u> try to cross it! But the 76-year old man did, and died.

Severity	Likelihood		
↓	Low	Medium	High
High	Medium	High	High
Medium	Low	Medium	High
Low	Low	Low	Medium

Fig. 11.4.-3 by 3 risk matrix, 3 categories

This does not mean death will happen every time, but the chance of it happening is very high, which must be avoided.

Any increase of categories beyond three should be dictated by the need to address – and the capability to control – a larger number of distinct categories.

That is, each category must have clear definitions and boundaries, with no gap or overlap between adjacent categories. There should be no ambiguity or doubt about the category for any particular combination.

Hence, if a company can define five distinct control groups for a 3×3 matrix, it would be proper to divide the risk zone into 5 risk categories, as in Fig. 11.5.

Note that the five categories for 3×3 are, fairly logically, along the five diagonals. The least and most risk zones are now one cell each.

At other extreme, too many categories, especially without clear and distinct definitions of each, would only lead to more paperwork and argument – and maybe even errors – without improving their utility.

Severity ↓	Likelihood		
	Low	Medium	High
High	Medium	High	Very High
Medium	Low	Medium	High
Low	Very Low	Low	Medium

Fig. 11.5. - 3 by 3 risk matrix, 5 categories

A practical maximum limit for categories is the number of diagonals in the risk matrix, which is ($m+n$–1), E.g. 6 for a 3 by 4 matrix.

Usually the number and boundaries of the risk categories are a matter of management policy and available funds, apart from need and capability.

Once the number of categories is decided, the boundaries may be chosen based on 'risk appetite', namely the level of risk the company prefers to 'not worry about' (acceptable) and the level it does not wish to face (unacceptable).

11.7. SIZE OF RISK MATRIX

The size of the risk matrix is simply the number of likelihood levels as columns (or rows) 'by' the number of severity rows (or columns).

With three levels as the minimum for both likelihood and severity, the minimum size of risk matrix is 3×3 (or, '3 by 3'), resulting in 9 combinations or 'cells' in the matrix.

The possibility of increasing likelihood and severity levels beyond three, if needed for special reasons, has already been discussed.

Some companies and government organizations in countries where they have long experience with risk matrices (and sufficient resources, of course) go to as big a size as 10×10.

A few even add a third dimension to the two already noted (likelihood and severity), such as 'exposure' to the hazard, and come up with a 10×10×10 risk solid, i.e. 1000 cells of 3D risk status!

The author wonders (with due respect) how assessors can decide between such fine divisions, except where all of them are precisely and quantitatively defined. Only the very rich with lots of time can afford it – often with doubtful validity of both input and outcome!

In sum, risk matrix size may be increased beyond 3×3 only when there is (a) the need to refine levels, and (b) the capability to implement controls.

11.8. THE CASE FOR 5×5

If we go beyond 3×3, what is the next good stop? By and large, it is widely accepted that an optimum size for the risk matrix is 5×5, corresponding to five levels of likelihood and severity, designated say 'Very Low', 'Low', 'Medium', 'High', and 'Very High'.

With a 5×5, we get 25 cells, and 9 diagonals. Table 11.1 gives specimen 5-level descriptions for likelihood and severity.

Table 11.1. Severity and Likelihood Levels for 5 by 5 Risk Matrix

| Risk Level | Likelihood Levels | Severity Levels of Impacts | | |
		Injury*	Environment Harm*	Loss
Very Low	Once in project lifetime or 10 yrs	First aid MC ≤ 1 day	Minor L T harm	< 5%
Low	Rarely, say once an year	Injury, medical treatment MC 2 to 7 days	Major L T or minor W T harm	5% to 20%
Medium	Occasionally, say once in 3 months	Serious injury, hospitalization MC > 1 week to 3 months	Major W T and minor L P harm	> 20% and < 35%
High	Frequently, say once a week	Multiple serious injuries, hospitalization Minor P disability MC > 3 months to 1 year	Major W T and minor W P Minor interruption to normal life and work	35% to 50%
Very High	Always or every day	Single or multiple fatalities Major P disability	Major W P harm Major interruption to normal life and work	>50%

*: L = Limited, W = Widespread, T = Temporary, P = Permanent

Note generally that it is fairly standard for participants in a course to evaluate instructors, patrons in a restaurant to rate various dishes, customers in a shop to judge the service, and so on, by marking one choice from a set of five: 'Very Bad', 'Bad', 'Average', 'Good', and 'Very Good'; or: 'Strongly Disagree', 'Disagree', 'Neutral', 'Agree', and 'Strongly Agree'; – or from some other set of five similar options.

Maybe 'five' has something to do with the number of fingers in each of our hands. (Remember that 'digit' actually means finger!)

Figure 11.6 gives a typical risk matrix for the 5×5 case, with the 25 cells shown grouped into 5 categories from 'Very Low' to 'Very High'.

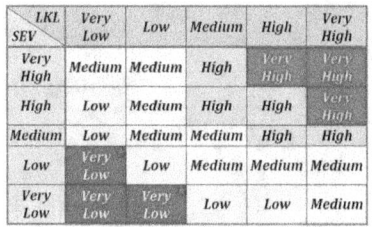

LKL SEV	Very Low	Low	Medium	High	Very High
Very High	Medium	Medium	High	Very High	Very High
High	Low	Medium	High	High	Very High
Medium	Low	Medium	Medium	High	High
Low	Very Low	Low	Medium	Medium	Medium
Very Low	Very Low	Very Low	Low	Low	Medium

Fig. 11.6. - Risk combinations for 5 by 5 matrix

More or fewer (but >2) categories may be used depending on what the company can define and control.

11.9. CHARACTERISTICS OF THE RISK MATRIX

(a) Square or rectangle:

A risk matrix need not be square. When numbers of likelihood and severity levels are not equal, the matrix will be rectangular.

(b) Symmetry of risk matrix:

For square risk matrices, usually the risk zone is taken to be symmetric, as in Figs. 11.1, 11.4, 11.5, and 11.6. The combination of 'Low' severity and 'Medium' likelihood is taken to be of the same risk as the combination of 'Medium' severity and 'Low' likelihood, namely 'Low' risk.

The same is true when every 'Low' term in the preceding statement is replaced with the term 'High'.

This kind of symmetry appears logical, and is easy to remember.

The same principle may be extended to rectangular matrices also, as in Fig. 11.7.

Severity	Likelihood				
↓	V.Lo	Lo	Med	Hi	V.Hi
Hi	Med	Med	Hi	Hi	Hi
Med	Lo	Med	Med	Med	Hi
Lo	Lo	Lo	Lo	Med	Med

Fig. 11.7.- Symmetry in rectangular risk matrix

(c) Contiguousness of categories:

Generally, cells in two adjacent rows or columns (and preferably along the diagonals also) will be the same or just one category away from each other, unless there is a special reason to skip a category. E.g. 'Low' should not be next to 'High', with 'Medium' missing.

It may be impossible sometimes to have contiguousness along all the diagonals (as in the basic 3×3 of Fig. 11.4) in which case, effort should be made to avoid a jump of more than one category.

It may be noticed that contiguousness is violated in Fig. 11.6 along two diagonals upwards from the star-marked cells, from 'Very Low' to 'Medium', skipping the 'Low' category. But it is maintained in Fig. 11.7.

This too may be avoided by re-designating the two star-marked cells as 'Low'. But this may be too drastic!

11.10. VARIATIONS IN RISK CATEGORIZATION

As already mentioned, risk matrices may be larger than 3×3. They may not be square. Also, more than three risk categories may be defined as long as the company can handle them.

Further, there may be occasions when symmetry is discarded in favor of practical considerations. Thus, referring to the 5×5 matrix of Fig. 11.6, a particular company may decide to deviate, and set the combination of 'Very High' severity and 'Low' likelihood to be 'Medium' risk rather than 'High' risk as marked.

To justify such asymmetry, reverting to the example of the man crossing a drain (Sec. 11.1), note (a) the 2 m deep and 1 m wide drain, and (b) the 1 m deep and 2 m wide drain, have both been marked as 'High' risk.

But if we analyze them more closely, in (a) the man is less likely to fall but he will be fully under water, while in (b) the man is more likely to fall, but he will be only waist deep in water.

The first combination is definitely more serious and less acceptable overall, than the second.

Relatively speaking therefore, we may revise (b) down to 'Medium' risk, at the extra bother of changing into a dry set of clothes if he falls!

Further, there may be situations when the category for a particular combination needs to be changed, and the original category is skipped.

Figure 11.8 shows an example from a Consulting firm

in USA. Its axes and sequence are different from our 'standard' – leading to most risky at bottom right. It is not square, and it is unsymmetrical. Finally, going from 'Low' in D-1 to 'High' in E-2, 'M' is skipped!

		Frequency					
		A	B	C	D	E	F
		Never Heard of Incident in EP Industry	Incident has Occured in EP Industry	Incident has Occured in SPDC	Happens Several Times per Year in SPDC	Happens Several Times per Year in Location	Happens every day
Consequence	1 Slight Injury	L	L	L	L	M	M
	2 Minor Injury	L	L	L	M	H	H
	3 Major Injury	L	L	M	M	H	H
	4 Single Fatality	L	M	M	H	H	VH
	5 Multiple Fatalities	M	M	H	H	VH	VH

Fig. 11.8. - 6 by 5 matrix from a U.S. company

After companies gain experience they may venture into these variations. But in every case, the authorities would insist on documented reasoning and supporting evidence to justify every variation from the basic, symmetric 3×3 matrix.

For our scope, we will carry on our discussion with the simple 3×3 matrix of Fig. 11.4.

11.11. LUMPING OF RISKS

The fact that there can be many different consequences whose severity can affect the well-being of a project and its personnel brings up another question.

Many companies who are currently doing hazard analysis assess various impacts of a particular hazard, and then take the average or the highest of the risks for further evaluation and control.

Some assign points to the different impacts, and then simply add the numbers to come up with an overall hazard severity level.

While this appears to be a logical approach, we must be careful that two consequences of different quality or criticality – such as serious physical injury and reputation damage – are not lumped together.

Unfortunately, if there is one critical impact (say injury) with a high numerical value, but all the other impacts happen to be low numbers, the total would remain below the threshold value chosen for serious risk control, and the hazard may get only little further attention.

This was precisely what happened in the recent case mentioned in Sec. 11.5.

There is also the fact that different types of consequences will require different controls, both qualitatively and in terms of expense.

For example, if in a job step, risk combinations from a physical consequence and a chemical consequence are respectively 'High' and 'Medium', it may be reasonable to rate the overall risk as 'High'.

However, when it comes to recommending controls, we may need to provide a high degree of control for the physical consequence but only a medium degree of control for the chemical consequence.

If the two are lumped together at start, then the common control will push the chemical part of it also to unnecessarily (and perhaps impossibly) higher levels than needed.

Alternatively, even though we may combine the risks into the highest level, we may subsequently choose to separate the controls to address the different consequences. In this case, any advantage gained by the earlier lumping together is lost.

Author therefore suggests that we retain the individual type of each hazard and consequence in separate rows right up to the controls stage – and not worry about the extra rows used up on the form!

Table 11.2 (showing only selected columns) illustrates the problem in lumping various consequences.

Table 11.2. Alternatives for Risk Control docmentation

Lumped Risks						
		Consequence and risk			*Overall risk*	*Con-trol*
Activity	*Hazard*	*Physical harm*	*Property damage*	*Reputa-tion damage*		
Spray painting at height	Falling from height	Major/ death [H]	Minor [L]	Adverse publicity [M]	Average M ? Max. H ?	?
Separated Risks						
Activity	*Hazard*	*Type:* Consequence	*Risk Level*	Control		
Spray painting at height	Falling from height	Physical harm: Major injury or death	H	Guard-rails, life-lines, anchors, safety harness		
		Property damage: Minor	L	(Monitor)		
		Reputation damage: Adverse publicity	M	Press meeting, fact sheet		

Author's suggestion to keep the consequence types separate is illustrated in the format of the Table's lower portion.

11.12. SEPARATION OF RISK LEVELS IN A JOB

It may occasionally happen that in a particular job, a first assessment results in all hazard likelihoods, severities, or risks as 'Low' and 'Medium', with no 'High' ones at all. Or all of them may be 'Medium' and 'High'. Or worse yet, all of them may end up at a single level.

But it would be quite rare indeed that all risks are at any single level – except possibly in a monastery! If such a thing happened, the RA team should review its boundaries of acceptability and unacceptability and fine tune its category assignments, to provide realistic risk separation.

In real life and from practical considerations, a risk assessment should display – and will naturally have – some 'Low', some 'Medium', and some 'High' risks the first time around.

To admit this, all risks should be evaluated at one time – or reviewed together – to sort out the acceptable and unacceptable risks, so that the tolerable risks are identified for further evaluation and control.

11.13. RISK MANAGEMENT IN PRACTICE – 6

"ALARP" stands for *"As Low as Reasonably Practicable"*. It is a particular mode of classification and management of risk, used in U.K. and some other countries.

The ALARP principle is represented as an inverted triangle, as in Fig. 11.9.

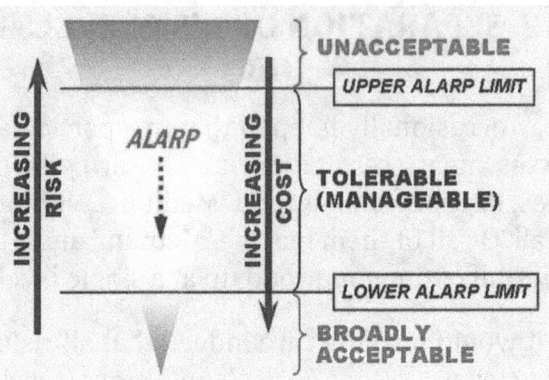

Fig. 11.9. - The ALARP triangle

Risk increases upward from bottom. Width of triangle at any height represents proportionately increasing expense of eliminating or mitigating the risk.

Although this RA method is used mostly for qualitative analysis and not likely to be considered an alternative to the procedure presented in this book, the concept of reducing residual risks to a level *"As Low as Reasonably Practicable"* is sound and all-encompassing.

"Reasonably practicable" involves weighing a risk against the trouble, time and money needed to control it.

Checking whether a risk is ALARP or not is a good step in self-assessment of an activity by a company. This will allow us to set goals for duty-holders, rather than being prescriptive.

———

12. NUMERICAL RISK ASSESSMENT

When the companies are big and the stakes are high, managers prefer quantitative assessments to qualitative ones, so that the controls can be planned more objectively and implemented more efficiently.

Quantification really means that every risk can be evaluated in absolute numerical terms for all the harm and loss, like so many dollars per year.

However, true quantification is very difficult and quite expensive. It involves massive data and resources, and the investment of considerable time, effort, expertise, and funds. The topic is beyond the scope of this book.

But a related technique is often mistaken for quantitative assessment, just because it involves numbers instead of words in the risk matrix.

The author wishes to present it as a useful alternative to the qualitative treatment presented up to this point, while at the same time demonstrating that it is not really quantitative.

12.1. FROM WORDS TO NUMBERS

Let us assign numbers 1, 2, ..., m to different likelihood levels, and 1, 2, ..., n to different severity levels, preferably starting with 1 for the lowest level (as is usual practice). *See* Fig. 12.1 for the (3×3) matrix.

Severity ↓	Likelihood		
	1 (=Low)	2 (=Medium)	3 (=High)
3 (=High)	3 (=Medium)	6 (=High)	9 (=High)
2 (= Medium)	2 (=Low)	4 (=Medium)	6 (=High)
1 (=Low)	1 (=Low)	2 (=Low)	3 (=Medium)

Fig. 12.1. - 3 by 3 Numerical Risk Matrix

Note that the numbers 1, 2, ... are simply identification tags – we may call them 'Ranks' – like the number on a footballer's jersey, and have nothing to do with the magnitude of the item.

The only conclusion we can draw – when we follow the convention of 1 being the lowest level – is that the larger the number, the higher the adverse effect, whether it be likelihood or severity.

It is for this reason that the method may be called "Pseudo-Quantitative", 'pseudo' meaning false or apparent, not real or actual. However, the simpler term 'Numerical' may be used, as long as it is remembered that the numbers are not values but ranks.

It is to be emphasized that in numerical RA (as against actual quantitative RA) only round numbers (integers) are used.

If more refinement of levels is needed, it may be tempting to use a numerical decimal value like 2.5 for a situation between 'Medium' (=2) and 'High' (=3), but that would be meaningless. What does 2.5 mean?

Obviously likelihood of 2 for something that happens

once a year is not just twice the likelihood of 1 for the same thing happening once a week; it is 52 times.

Likewise, severity of 3 for death is not just 3 times the severity of 1 for a scratch, but more like hundreds of thousands of times.

Hence, if there are cases which fall between 'Medium' and 'High' and we want to classify them separately, we would have to increase the number of levels m and/or n.

We may call these in-betweens as 'High' (= 3), and add another row or column for what we had originally called 'High', re-naming them 'Very High', and marking their rank as '4', as shown in Fig. 12.2.

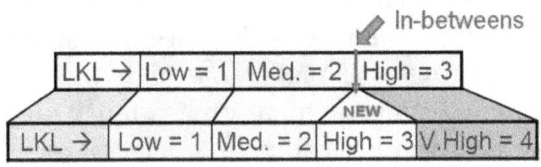

Fig. 12.2. - Insertion of extra level between existing ones

12.2. FROM NUMBERS TO ARITHMETIC

Once numbers are assigned to identify likelihood and severity levels, they may be manipulated in some fashion to represent the combined effect, giving numerical 'risk indices'.

One may add the likelihood number and the severity number; or one may multiply them. The results will of course be very different.

For instance, in a 3×3 matrix, the most risky value will be (3+3) i.e. 6 by addition, but (3×3) i.e. 9 by multiplication. This variation in risk value will definitely cause confusion in the reporting of risk indices.

Actually, there is a big body of scientific work behind the mathematical determination of risk indices from numerical likelihood and severity values. In these, addition and multiplication do have quite different implications, discussions of which are beyond the scope of this book.

For our present purpose, let it suffice to say that under normal conditions analyzed in practice, multiplication is more appropriate than addition. Unless otherwise specified, multiplication will be assumed.

When the author comes upon any company using addition, he asks them why they are adding instead of multiplying. If they are unable to explain their rationale, he urges them to change to multiplication, if only so that their new staff or other companies may not misinterpret their results.

Accepting multiplication as the norm here, cells in the risk zone should be filled with the products of the likelihood and severity level numbers:

Risk index = (Severity Level No.) × (Likelihood Level No.)

Risk Index is also termed 'Risk Prioritization Number' and designated 'RPN"

There will be ($m.n$) cells in a matrix with m columns and n rows, and they will be filled with numbers 1, 2, ... ($m.n$). The lowest risk index will be 1, and the highest risk index will be ($m.n$).

The numbers in each row (and column) may be obtained by multiplying the likelihood level number of each column with the severity level number of the particular row (and multiplying the severity level number of each row with the likelihood level number of the particular column), getting, 1, 2, 3, ...; 2, 4, 6 ...; 3, 6, 9, ... and so on.

Note from Fig. 12.1 that because of the restriction to integers for the likelihood and severity levels, some risk indices (2, 3, and 6) will be repeated, and some intermediate numbers (5, 7, and 8, for the 3 by 3 matrix) will be missing.

Obviously, the risk indices are not to be taken as absolute or even relative values. For instance, a risk with an index of 2 will certainly not be just twice as risky as one with an index of 1, but much, much more.

Again, as long as we start with 1 for the lowest likelihood and severity levels, the larger the risk index, the higher will be the risk. That is a main psychological benefit from the recommended numerical scheme.

A clear advantage of the numerical method over the qualitative is that in the numerical method, a secondary prioritization may be made within the same qualitative category. For instance, within the medium risk category, risks with index 4 may be given higher priority than those with index 3.

Symmetry of risk indices is automatically maintained. Risk with severity 2 and likelihood 1 will be the same as risk with severity 1 and likelihood 2, as both result in a risk index product of 2.

While this outcome may not be valid for all cases, it is

alright for starters and may apply to most common situations. Any desired asymmetry as described in Sec. 11.9 would be difficult to implement with the numerical scheme in which symmetry is automatic. Further, if and when implemented, the hazard and its control will have to be monitored and followed up as an exception.

12.3. NUMERICAL RISK CATEGORY ASSIGNMENT

The risk indices will again have to be grouped into a certain number of risk categories, for each of which, controls and procedures must be specified. A minimum of three categories is essential, as already stated.

For three categories, risk indices of Fig. 12.1 may be grouped as shown in the left block of Fig. 12.3.

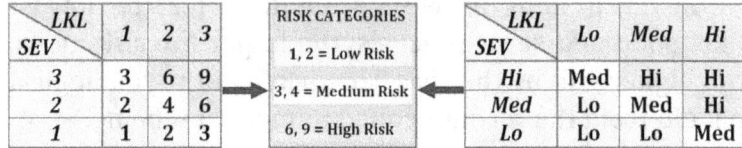

Fig. 12.3. - Numerical risk categorization

Before we can use the numerical risk matrix, we must group the risk cells into various categories just like for the qualitative risk matrix, as shown in the right block of Fig. 12.3, which is the same as Fig. 11.4.

As we need both qualitative and numerical risk matrices to lead to the same categorization, we must declare the equivalence through the listing of numbers for various risk categories as shown in the middle block of Fig. 12.3.

Now, based on the risk categories 'Low', 'Medium',

'High' etc., appropriate controls may be recommended and implemented. The various categories must be described clearly and completely.

A similar procedure may be applied for a 5×5 matrix, as shown in Fig. 12.4. On comparison with Fig. 11.6, it may be noted that Fig. 12.4 satisfies contiguousness fully while Fig. 11.6 did not.

LKL SEV	1 (=VL)	2 (=L)	3 (=M)	4 (=H)	5 (=VH)
5 (=VH)	5	10	15	20	25
4 (=H)	4	8	12	16	20
3 (=M)	3	6	9	12	15
2 (=L)	2	4	6	8	10
1 (=VL)	1	2	3	4	5

V = Very; L = Low; M = Medium, H = High
RISK CATEGORIZATION
1 = Very Low, 2-4 = Low, 5-10 = Medium,
12-16 = High, 20-25 = Very High

Fig. 12.4. - Categorization of 5 by 5 risk matrix

Thus, whether we use qualitative or numerical method, final decisions on risk categories are still made only on how the RA team and management plan to handle the various combinations of likelihood and severity. Remember, these are not formulas to apply blindly!

The capability of sub-prioritization within a qualitative category has been discussed in Sec. 12.2. Apart from this benefit to decision-making, there is no intrinsic advantage of one method over the other. However, we cannot, and should not, (a) do any further mathematical juggling with the numbers, or (b) draw quantitative inferences from them.

Additionally, there are human factor advantages with numbers. When and where language is a problem, numbers speak louder and clearer than words. It is also a fact that most engineers and technicians are more comfortable with numbers than with words.

The use of numbers overcomes the vagueness of words, especially when actual users of risk matrix are not sufficiently educated or trained. Numbers also make communication and evaluation easier within a project. Once validated, further application becomes simple and fast.

Instead of explaining a 3×3 qualitative matrix to a supervisor (and his explaining to a foreman) that *"combination of high severity and medium likelihood leads to high risk"*, it is enough to say that *"a risk index of 6 (got from 3×2) is high"*.

It is easier to communicate the modifications to the line personnel and enable them to apply the revised guidelines with numbers rather than with words.

12.4. RISK MANAGEMENT IN PRACTICE – 7

22 January 2007 OSH Alert
Occupational Safety and Health Alert

Series of Forklift Accidents

In 2006, there was a spate of (FOUR) fatal accidents involving forklifts that, sadly, could have been prevented.

Tip: Have you conducted your risk assessment? The Workplace Safety and Health (Risk Management) Regulations require all workplaces to conduct risk assessments to identify the source of risks, actions that should be taken and parties responsible for doing so.

The point to note here is that conducting RA is a mandatory requirement of safety management in many countries.

In Singapore, failure to conduct RA is punishable with a SG$10,000 for first offence, and double the fine and/or 6 months imprisonment for succeeding violations.

---

13. RISK CONTROL CONCEPTS

13.1. PRINCIPLES OF RISK MANAGEMENT

(a) Accept risk only when benefit exceeds risk.

Risk is inherent in the nature of many industries.

Risk is usually proportional to gain.

All risk cannot be eliminated.

(b) Accept no unnecessary risk.

An unnecessary risk is any risk that, if taken, will not contribute meaningfully to mission accomplishment.

Managers who accept unnecessary risks are gambling with the lives of their workers. The gambler does not know what will happen. Only a good risk-manager can reasonably predict it.

(c) Anticipate and manage risks by planning.

Risks are more easily controlled when identified at the planning and design stages.

If risk controls are tacked on as an afterthought in training or in the execution stage, they will cost too much and probably fail.

(d) Eliminate or reduce risks at source.

The WSH Act aims to do this by making stakeholders accountable for managing the risks they create.

For example, exhaust toxic fumes in electroplating at the bath level where they start, rather than by a hood near ceiling and gas masks for workers.

(e) Make risk decisions at the right level.

The RA team makes recommendations and the management approves them. Management and RA team set likelihood and severity levels and risk categories.

The manager directly responsible for a job should understand and accept the recommended risk decisions.

If and when risk exceeds the benefit during an operation, or help is needed to implement controls, communicate and consult with higher authority.

13.2. WHAT TO DO WITH THE RISKS

As we would have already included the existing/required controls for the job, the further controls after RA would be 'Additional Controls'.

To keep the subsequent discussion simple, regardless of the number of levels the likelihood and severity, and the number of categories of risk, we shall refer to the lowest category as 'Low', the highest category as 'High', and everything in between as 'Medium'.

In Fig. 13.1, author depicts the three risk zones as the 'Medium' or tolerable risk space bounded below by a floor (upper boundary of 'Low' risk space) and above by a ceiling (lower boundary of 'High' risk space).

The tolerable risk space between the floor and the ceiling in the figure is the zone in which the risks must mainly be managed on an everyday basis. This is 'living space' which must be continuously monitored and managed.

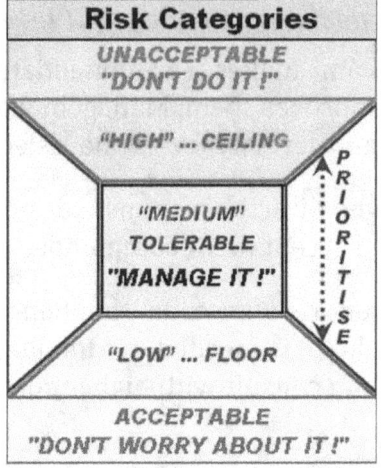

Fig. 13.1. - Risk space and categories

[*See* also 'ALARP' in 'Risk Management in Practice – 6' at the end of Chapter 11.]

The author's *'manthra'* (meaning slogan, chant) for the three zones are also marked.

(a) Low Risk – "Don't worry about it!":

'Don't worry about it!' does not mean *'Ignore it'*.

It simply means that the company need not invest money and time to control it, under current circumstances.

At the same time, supervisors and foremen need to keep an eye and an ear out to monitor 'Low' risk tasks, simply to make sure that none of the identified low risk hazards gets worse and begins to harm people or cause damage.

Just like a mother who leaves her child playing in the

living room while she is busy in the kitchen: She doesn't worry as long as the child is occasionally making burbling noises.

If there is silence for too long or if the mother hears a thud or a scream, she immediately knows something is wrong and rushes to find out what it is!

(b) High Risk – "Don't do it!":

'Don't do it!' means exactly that – just DO NOT do it!

When a company's RA team and management have themselves evaluated a job step as of highest risk, and hence 'Unacceptable', it wouldn't make sense to go ahead and do it.

The pressure to change a risk from highest category to something lower will be very high. It may even become a matter of survival of the company if some fresh tasks or some ongoing tasks turn out to be highest risk, thus requiring immediate stoppage of the job.

In such cases, the company must not start the new job. If it is already doing a high risk job, it must stop the job immediately.

Then, it must do something different and something more to improve safety of the task – like adding more supports, getting a more powerful machine, employing more qualified personnel, etc. – and re-do the RA.

Hopefully the risk would have improved from the highest risk level to next lower level which would allow the company to proceed with the task, with additional safeguards.

If the company's resources (or time) do not permit

such an improvement, sub-contracting the 'High' risk task to a specialist firm would be a wise move for overall effectiveness and safety.

(c) Medium Risk – "Manage it!":

The 'Medium' tolerable risks should be prioritized on some basis, first by physical harm and next by cost, effort, and time factors.

Table 13.1 gives general recommended control actions for three risk categories.

Table 13.1. Risk controls for 3 categories

Risk category	Recommended actions
Low (Acceptable) [Don't worry!]	No additional risk control measures needed. But frequent monitoring needed to ensure that the risk level assigned does not worsen over time.
Medium (Tolerable) [Manage it!]	Risks should be prioritised and managed so as to reduce risk levels to as low as is practicable and maintain them there, within specified time periods. Interim risk control measures needed. Management attention and approval required.
High (Unacceptable) [Don't do it!]	Job must not be started. If work is going on, it must be stopped immediately. Eliminate or reduce to at least 'Medium' risk before commencing or resuming work. No interim control measures. Risk controls not to be too dependent on PPE.

Phrases in square brackets are Author's control 'Manthras'.

Table 13.2 gives a similar set of recommendations for five risk categories.

Table 13.2. Risk controls for 5 categories

Risk Category	Recommended actions
Very Low ('*Trivial*')	No additional risk control measures needed. Monitoring needed to ensure that risk does now worsen over time.
Low ('*Limited*')	Document and monitor. No urgent action required. Limited and affordable remedial action as and when necessary.
Medium ('*Moderate*')	Check, modify, and proceed. Early action required. Additional (administrative) controls; fair amount of resources needed.
High ('*Substantial*')	Stop operation, rectify, and continue; urgent action required. Additional (engineering) controls; substantial resources needed. Work suspended or restricted to interim controls.
Very High ('*Catastrophic*')	Job must not be started. If it is going on, work to be stopped immediately. Eliminate or reduce to at least 'Medium' risk before commencing or resuming work. No interim risk control measures. Risk controls not be too PPE dependent.
Category names in single quotes are sample names used by some companies.	

This Medium risk region may also be further divided. The first and last rows of Tables 13.1 and 13.2 are the same, and only the middle row of Table 13.1 has been subdivided into three rows in Table 13.2.

All medium risks must be managed by one or more control methods according to a conventional hierarchy, presented in next chapter.

13.3. BUSINESS CASE FOR RISK MANAGEMENT

Are all these risk controls worth the effort and the cost – apart from the many benefits from RA and RM (as discussed in Section 3.5)?

Yes, the author has already mentioned the twin value of money in RM:

1. Money is the language of business, even of government at some level. It is the measure to make effective decisions pragmatically.

2. Even more important, in today's context, investment in risk assessment and safety management is actually a very profitable venture. Benefits can be shown to outstrip costs many times over.

The key to the business case is the fact that the direct, visible cost of accidents is but a fraction of the total costs, with the major indirect and unknown costs remaining hidden, just as the tip of an iceberg (Fig. 13.2) above the water is only a small part of its total mass.

The bulk remains hidden under the ocean surface, ready to destroy the unwary 'Titanic'-s of even very big companies.

The ratio of hidden costs to direct costs vary from about 20 for small expenses to about 1.5 for very large expenses, average being about 5 .These usually show up after accidents.

For example, for an eye injury, the hospital costs would be about $5000, hidden costs may be say 8 times as much. To recover this total of $45,000 at a profit

margin of 3%, the company must generate (45000×
100/3) i.e. $1,500,000.

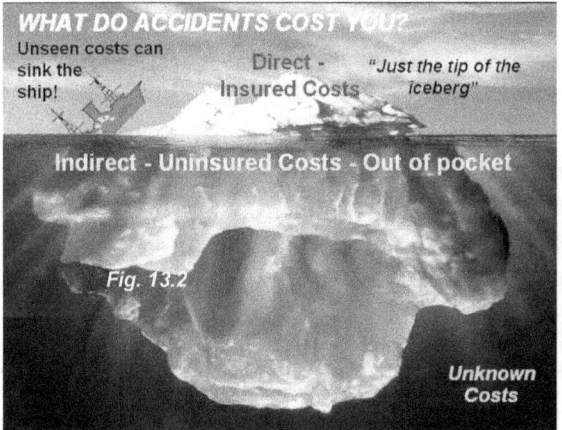

Fig. 13.2. - Iceberg Analogy for hidden costs of accidents

All for lack of a pair of goggles costing $10!

On top of the immense financial savings from proper
risk management will come other tangible and intangible
benefits also.

———

14. RISK CONTROL METHODS

14.1. RISK CONTROL HIERARCHY

Unacceptable high risks must be reduced to at least medium, and acceptable low risks must be monitored so they do not worsen. Tolerable medium risks must be managed to retain them at the medium level and if possible to reduce them to low level.

All these goals require some control action. Risk management follows a conventional hierarchy of controls, the word 'hierarchy' referring to the decreasing order of effectiveness of the control action, usually as follows:

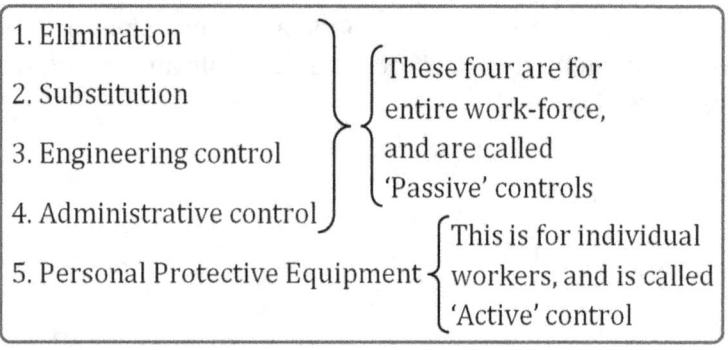

1. Elimination
2. Substitution
3. Engineering control
4. Administrative control

These four are for entire work-force, and are called 'Passive' controls

5. Personal Protective Equipment

This is for individual workers, and is called 'Active' control

This may be shown pictorially also as in Fig. 14.1.

14.2. ELIMINATION

First option, the most direct choice, is to eliminate the hazard. This is a final solution, because after this the hazard itself is not in the list, and need not be considered any further.

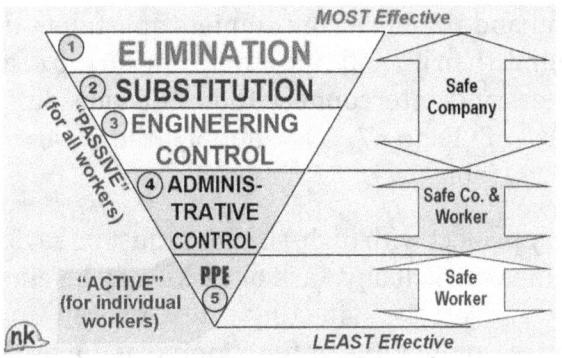

Fig. 14.1. - Hierarchy of risk controls

In fact, in our personal life, we do this instinctively, even to excess. Most don't gamble because they cannot afford to lose. Likewise: If you cannot swim, don't go near the water.

- If heights are a problem, don't climb up – or down.

- To reduce dengue attacks, eliminate mosquitoes. (Singapore has successfully reduced dengue deaths, by intensive campaigning and preventive action.)

- To eliminate the fatal risk of fall from heights, just eliminate scaffolds.

But elimination is not always feasible or cost-effective.

Americans say: *"If you can't stand the heat, get out of the kitchen!"* But if we do not want to go out to food stalls all the time or we cannot go out for some reason, we have to go into the kitchen to cook and eat.

In a similar vein, is elimination of scaffolding possible in construction?

In certain cases, it is possible.

Japan and UK fabricate complete apartment units on the ground, then lift and stack them one on top of another by cranes, and interconnect them. Canada did it with 'Habitat' in Expo'67. No doubt, these are highly specialized jobs.

Every project with high-risk jobs must be reviewed to determine which, if any, task or tasks may be eliminated without adversely affecting the overall purpose, economics, completion, or functioning of the project.

Even transferring a high-risk task to specialists amounts to eliminating it from the current scope.

14.3. SUBSTITUTION

Substitution of a hazardous material, equipment, product, process, or person, with a less risky one, is a feasible control option.

We do this also spontaneously in our personal life. Rather than scramble up or down too steep (or too wet) a slope, we go around.

At a workplace, we should:

- Break down heavy loads into lighter portions.

- Driving on a freeway at a normal 90 kph (i.e. 56 mph) speed, when it starts raining, cut the speed down to (i.e. substitute with) a smaller speed (say, 45 kph, i.e. 28 mph).

- For site work exposed to rain or other wetness,

instead of electrically operated tools, use compressed air for the power.

- Instead of toxic solvent-based paints use water-based paints.

- Use liquid instead of gas, powder instead of liquid, and solid instead of powder, when toxic chemicals are involved – recognizing a reasonable increase in price.

- Reduce pressure, temperature, and other hazardous variables to the minimum, or find alternatives.

- Switch from 220 v to less dangerous 110 v for site tools.

14.4. ENGINEERING CONTROL

This will constitute the bulk of risk management applications to engineering and technology in actual practice.

This includes, but is not limited to, the following:

- Guard-rails and toe-boards
- Noise baffles
- Automatic fire sprinklers
- Chemical filters
- Earth Leakage Circuit Breakers
- Mechanical aids
- Trolleys for heavy loads
- Exhaust hoods
- Machine guards, lock-outs

- Anchors and lifelines
- Rescue equipment
- Safety nets
- Fire extinguishers
- Alarms, sirens etc.

As engineering controls are the most common practical measure for safeguards in industries, a general recommendation here is to have some redundancy built into the system, that is, to provide at least one extra back-up control, to ensure effectiveness.

An example would be the visual and sound alarms inside crane cabins for over-reaching or overloading.

14.5. ADMINISTRATIVE CONTROL

This is the direct contribution of the management. Apart from promoting policy and funneling funds towards the cause, the administration will be responsible for the following tangible activities:

- Support and empower the RA team
- Facilitate safety meetings, suggestions, violation reports, etc.
- Provide warnings, markings, placards, signs, and notices
- Provide tag-outs for dangerous equipment / situations
- Develop, monitor, and implement written policies, programmes, instructions, and SWPs
- Obtain approvals, permits to work, etc.

- Limit number of personnel/equipment exposed to hazards

- Limit frequency, duration, and severity of exposure to hazards

- Provide contingency funds for urgent safety-related expenses

- Grant authority to supervisors and foremen to suspend unsafe workers, change fatigued workers etc.

- Facilitate certification and periodic safety training to staff

- Ensure proper and strict inspection and supervision, particularly in the implementation, maintenance, and use of all safeguards

These are usually indirect and supplementary controls to facilitate and enable other controls, rather than direct control of hazards. Administrative controls by themselves cannot effectively reduce risk.

14.6. PERSONAL PROTECTIVE EQUIPMENT (PPE)

The preceding four controls apply to the entire workforce, and do not require the worker to decide or do anything on his own except to follow the relevant rules and good practice. They are called 'passive' controls.

But PPE, the last of the controls and assigned the lowest priority, applies to the individual worker 'personally'. It is an 'active' control, as it is the worker who must understand, accept, and adopt it.

Common PPE include:
- Helmets
- Goggles
- Hand gloves
- Steel-toed boots
- Ear plugs
- Gas masks
- Protective overalls
- Waist belts
- Full body harness

This is the least effective control, the weakest link in the chain of safety implementation, because:

- We need as many pieces of the PPE as there are workers.

- The individual worker may not use the PPE or use it wrongly, out of ignorance, negligence, over-confidence ('macho' type bravado), anger, fatigue, boredom, or other reasons.

- Unlike the other controls, PPE is worker specific, namely sizes and fittings must correspond to the individual.

- The worker himself, or some specific person in the loop – supervisor in particular – must take care of regular maintenance of the PPE.

- In any case, an employer cannot treat PPE as unimportant.

It is the last line of defense for the worker, if and when everything else fails.

14.7. INCREASED RISK FROM CONTROLS

A couple of critical outcomes of all these additional safety measures that can occur occasionally must not be overlooked:

1. The PPE itself may introduce fresh risk. E.g. Ear-plugs may prevent a worker from hearing verbal or other sound warnings or instructions.

2. Elimination or substitution of one risk may introduce another risk.

3. Reduction of likelihood may increase the severity, or vice versa. E.g. Removal of a traffic visual block at a blind corner may result in fewer accidents but these may be of greater severity because of the higher speeds.

4. Too much PPE or PPE systems with missing co-requisites may introduce fresh hazards. E.g. Body harnesses without sufficient fall clearance or adequate anchor may result in serious injuries.

14.8. OUR MAN CROSSING THE DRAIN

Our man, still thinking about crossing the drain, may be faced with the following risk control options – four provided by the authorities or owners, and one by himself, as depicted in Fig. 14.2:

1. Elimination: The drain is covered up, after embedding a pipe to carry the water. (Or just drop the plan to cross!)

2. Substitution: A bridge has been built, as a much less

dangerous alternative to crossing the drain. Even with this option, some, like the man in the news, would want to take a short-cut!

Fig. 14.2. - Risk controls for crossing drain

3. <u>Engineering controls:</u> Barriers along both sides of the drain.

4. <u>Administrative controls:</u> Funding for the controls, placement of warning signs.

5. <u>PPE:</u> Full body water immersion suit, life belt, and snorkel. [*Of course, I jest – I am forced to say this because one of my students took me literally! – Author*]

14.9. WHAT WILL THE CONTROLS HELP IMPROVE?

This question was raised in Section 8.2 on existing controls.

Controls are intended to reduce likelihood and/or severity. As long as one or the other factor is reduced, the resulting risk should also reduce.

But whether likelihood is reduced or severity is reduced depends on the particular control. Author sorts the two groups of controls as follows:

(a) Type 'L' control:

If the control prevents the hazard from becoming a risk, that is, if it is aimed at eliminating or preventing the accident, and the worker is not exposed to the particular hazard (or less exposed to it than before), then it reduces the likelihood of the hazard.

This is 'pre-mishap', and becomes effective to prevent the mishap and/or reduce its frequency and exposure.

(b) Type 'S' control:

If the control cannot prevent the mishap (that is the worker cannot be prevented from exposure to the particular hazard) but it is aimed at reducing the harmful effects of the hazard if, when, and after the hazard is realized, then it reduces the severity of the hazard.

This is 'post-mishap', and becomes effective only after the mishap occurs, serving to mitigate the bad effects of the mishap.

(c) Type 'LS' control:

There may be Type 'LS' control too, combining both beneficial effects of a single control reducing likelihood and severity. If this happens, so much the better!

14.10. IMPROVEMENTS BY HIERARCHY

We will analyze all five controls in the hierarchy, using this categorization:

(a) Elimination:

Once a hazard is eliminated at source, it will have no further effect on the job, and there is no question of which factor it will reduce.

(b) Substitution:

This is a fairly common control. Substitution of a product, process, or person by a less hazardous alternative can reduce severity and/or likelihood.

Type 'L' Examples: Paint application by brush instead of spray; use of solid chemicals instead of liquids or gases.

Type 'S' Examples: Using water-based instead of solvent-based paint; mechanical power instead of muscle power.

Type 'LS' Examples: Reducing speed from 90 kph to 45 kph during storm. The driver will have twice the time to notice an oncoming car or other obstacle and try to avoid the accident, reducing likelihood by half.

Even if there is an accident, the resulting energy transfer will be reduced to one-fourth, and hence severity will be much reduced, from death to fractures, from fracture to bruise – from totaled car to fender-bender.

Some vaccinations, which reduce both the likelihood of the infection as well as the severity if and when it happens.

(c) Engineering controls:

These common safety measures may be aimed at reducing the likelihood and/or severity of the hazard to the worker. Most of the substitution controls also depend on major contributions from engineering.

Type 'L' Examples: Guard-rails prevent worker from exposure to fall; machine guards prevent contact with moving parts; face masks reduce likelihood of catching or passing on a cold.

Type 'S' Examples: A net, mat, or air-cushion under or around a work platform reduces the severity of harm if and when the worker falls.

(d) Administrative controls:

Type 'L' Examples: Warning signs, brochures, posters.

Type 'S' Examples: First-aid training.

Type 'LS' examples: Worker and supervisor training and certification; SWP, worker rotation.

(e) Personal Protective Equipment:

This is the last resort, for use when nothing else can be guaranteed to work or as a supplemental safeguard on an individual basis.

Type 'L' Examples: Radiation detection badge, waist-belt and work-restraint cable. These are warning and preventive safeguards to avoid mishaps to individuals.

Type 'S' Examples: Helmet, gloves, special shoes, goggles, ear-plugs, full-body harness. These more common PPE are effective only after the mishap, and only if and when the worker wears the equipment and uses it correctly. These will not reduce the likelihood of occurrence, but will reduce the severity of consequence of the mishap.

All preceding points apply equally to existing and additional controls.

14.11. ADDITIONAL CONTROLS

Most of the time, from common sense and self or shared experience, the risk assessment of a typical workplace job will be safe under existing controls.

But in a complex or new job, there could be some activities which have some high risks or too many medium risks.

The high risks have first to be brought down to at least medium. Attempts must also be made to bring the medium risks to low, and to keep the low risks from becoming worse. For all these desirable outcomes additional controls must be provided according to the hierarchy listed.

Once an additional control has been selected, a person must be assigned to implement it either by name or designation – or specifically referred to a division, together with a date or a timeframe by which to complete the implementation.

14.12. CAN ALL RISKS BE BROUGHT DOWN TO 'LOW'?

Such a result would be practically impossible. It would be like reducing a construction site activity to the same level of watching TV at home.

We must accept the ground realities: Many industrial

undertakings involve some degree of risk. A common example is the risk of falling from height. Most safeguards would only reduce the likelihood, leaving the severity 'High', so that the residual risk would be 'Medium' at best.

Demonstrating that we have reduced all risks to acceptably low levels may satisfy our egos and maybe some casual inspection. But the effort, cost, time and responsibility to get them all to 'Low' will be more overwhelming than if we declared the existence of some 'Medium' level risks which must be continually managed by further practical and reasonable measures.

The assessor must be clear of the goal of the exercise: Whatever logic or name-tags are used, whether controls are in one column of the risk assessment form or another, at the end of the exercise the residual risks must be within the safe ('Medium' or 'Low') domain. That is all.

It would be wishful thinking, and in fact, as the psychologists would say, downright 'denial' (meaning fooling ourselves), to show that we have eliminated and/or reduced all the risks to acceptable ('Low') level.

This might be possible in theory and on paper, by repeatedly adding more and more controls until both likelihood and severity dropped finally to 'Low'. However, there would be two problems with such a move:

1. Users and inspectors would then be lulled into over-confidence and a false sense of security, and stop worrying or even thinking about the lurking hazards. Before long, a big accident might happen.

2. The point of diminishing returns, namely where the

cost begins to exceed the benefit, would be crossed long before all the risks could be lowered sufficiently. This usually serves as a self-correcting mechanism against the desire to lower all risks to acceptable levels.

Thus, there would – and author feels there should, to prevent complacency and neglect – always be some residual 'Medium' risk to monitor and manage.

If a sick child or an old man can be cured by medicine, still wouldn't it be a good idea for a responsible person to see that the patient takes it?

14.13. COMPLIANCE REGIME

The 'L' and "S' Types of all five hierarchical controls described earlier, whether existing or additional, are specific product- or process-related events and practices, for implementation once, or at specified intervals.

How then, can the staff 'manage' the remaining medium residual risks after all the feasible additional controls have been implemented?

The answer is, by a strict and comprehensive regime of the following:

1. Maintenance, of plant, PPE, etc., including repair
2. Inspection, of all equipment and PPE, SWP implementation, etc.
3. Supervision, of all activities and processes

Maintenance and inspection are done at regular intervals ('planned') or as condition-based ('on-demand'). Supervision should go on all the time, especially for hazardous tasks.

Author will refer to this triple requirement in risk management as 'C-Regime' for short – 'C' for 'Compliance' (and for 'Continuing', in the sense of throughout the project, and until its completion.)

In contrast to the other controls, the characteristic of the 'C' Regime is that all three of its aspects depend on positive action by, and accountability of, responsible and competent regular staff throughout the project and routinely, to ensure the effective functioning of all specified safeguards.

Compliance regime is a pre-requisite to risk management, because without any one of its three components, all safeguards would be ineffective – just as if the patient simply did not take his medicine!

All concerned staff shall be trained for, familiar with, and committed to, carrying out relevant duties as per their contracts and SWPs, with due diligence.

Some may list 'C' regime items in the RA form simply as a reminder to the concerned staff. As they would apply to all the safeguards in the last four of the five controls – not being required for the eliminated risks – such a reminder would then have to be listed for every proposed control, as otherwise it may send a wrong message that the regime is not critical where it is not mentioned.

Instead, at points along the chain of command, the importance of the C-regime shall be emphasized to all the responsible staff (from project managers to workers) via Safe Work Procedures and method statements. They should be routinely required to include in their detailed reports, specific references to the C-regime for every medium risk task.

Some risk assessors include the C-regime under 'Additional Controls' to reduce 'Medium' risk levels to 'Low'. But, for reasons mentioned earlier, author prefers to keep it separate from the standard controls. In fact, Singapore specifically discourages personnel from listing these three mandatory requirements in their risk assessment forms.

The recommended procedure is to consider the compliance regime as part of the professional 'Duty of Care', and presume its existence in every case as a mandatory requirement.

The only time the C-regime needs to be included in the risk assessment form may be when a new piece of equipment or a new procedure is introduced into the system, or when a new foreman is assigned or a new or extra worker is brought in.

In such a case, 'Additional Controls' may highlight the particular special and non-routine C-regime activity, for a specified duration only.

---

15. USES OF RISK MATRIX

Risk matrix is basically just a table, a graphical device, displaying risk categories for various combinations of likelihood and severity levels.

From this, the risk category may be picked off for a job step with given or known likelihood and severity levels.

However, risk matrices have many other uses, some of which are given below. For illustration purposes, let us consider a job with 9 steps, in a risk zone with 5 categories.

The risks associated with the nine steps may be tabulated as shown in Table 15.1 along with the risk matrix used, namely the same one displayed in Fig. 11.5.

Table 15.1. Example of use of risk matrix

Step No.	Likeli-hood	Seve-rity	Risk category
1	L	L	Very Low
2	L	M	Low
3	M	M	Medium
4	M	H	High
5	H	H	Very High
6	M	M	Medium
7	H	L	Medium
8	H	H	Very High
9	L	L	Very Low
L = Low, M = Medium, H = High			

Severity ⬇	Likelihood		
	Low	Medium	High
High	Medium	High	Very High
Medium	Low	Medium	High
Low	Very Low	Low	Medium

15.1. STATUS MAP

A very simple and effective use of the risk matrix is as a map of the locations of the various risks within the appropriate cells, so that the matrix now serves as a visual display of the risk status of all the nine steps of the job as in Fig. 15.1.

Severity	Likelihood		
↓	Low	Medium	High
High		④ M	⑤ ⑧ VH
	M	H	
Medium	②	③ ⑥	
	L	M	H
Low	① ⑨		⑦
	VL	L	M

Fig. 15.1. - Risk matrix as status map

Although the table gives the same information, the risk matrix status map visually displays the risk distribution from very low to very high.

It highlights in particular that the company must immediately do something about steps 5 and 8, with unacceptable very high risks, as without that, it cannot proceed with the job. It also shows that step 4 has to be watched carefully and its risk reduced on a priority basis.

15.2. RISK PRIORITIZATION

One of the primary uses of the risk tabulation is the capability to sub-prioritize within the same category.

In qualitative matrices, organisations may choose to differentiate risk levels from combinations of the same two levels of severity and likelihood. Often, considering severity as more critical than likelihood, combination of medium severity and low likelihood may be taken to be more risky than the combination of low severity and medium likelihood. They must highlight it – and even explain the distinction – to their users in the risk matrix.

Advantage of another kind of sub-prioritization automatically occurs in numerical risk matrices such as in Fig. 12.1 and 12.4, which are automatically symmetric.

To illustrate: In the 3 by 3 risk matrix of Fig. 12.1, combinations of (L=1, S= 3), (L=2, S=2), and (L=3, S=1), giving 3, 4, and 3, are treated as 'Medium' risk. However, as 4 is larger than 3, the second of the combinations may be treated as more critical than the first and the third.

15.3. DECISION MAKING

The map is also a decision-making guide. The general aim would be to shift the risk status of all the job steps horizontally and/or vertically, corresponding respectively to likelihood and severity, from the higher level risks continually towards the least risk corner.

With the scatter visually available, the company can think and plan various scenarios for improvement.

For instance, what can the company do to move steps 5 and 8 out of the unacceptable region?

Perhaps it does not have the resources (or the time) to shift step 8. It may farm that task out (that is, sub-contract it), to experts in the task who have the resources and experience to carry it out smoothly and safely. Once step 8 has been farmed out, it will not appear any more in the list of the company's risks.

For step 5, the company may be able to apply additional controls to shift its position leftwards or downwards from the top right corner to the cell to the left of it, moving it away from the unacceptable very high risk zone to a tolerable high risk zone, as shown in Fig. 15.2. We may now cross out both steps 8 and 5, the former shown moving out, and latter shown moving left.

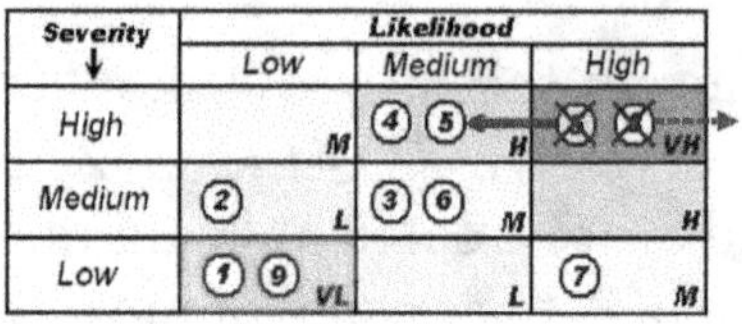

Fig. 15.2. - *Managing very high risk steps*

With the additional safeguards and approval of the authorities, the company may now carry on the work including Step 5.

15.4. CHANGES IN RISK STATUS

The chart may be used to record improvement, deterioration, or other change in status from one assessment to the next.

It must be realized that the risk matrix is not a one-time exercise to be wrapped up and put away for the mandatory maximum interval, or till the next accident. Almost everything in the workplace is in dynamic flux, and most things deteriorate with time. That is why most risk management codes recommend a fairly regular monitoring of the risk status, Singapore going so far as to suggest monthly reviews.

In this case, let us say over a period of time, some equipment or process has deteriorated to causing increasing harm to users, raising the corresponding step (say No. 9) from 'Low' severity to 'Medium' severity, as shown in Fig. 15.3. (Note that now the farmed out step 8 is omitted and step 5 is showed in the moved position.)

Then the revised risk level of step 9 climbs from 'Very Low' in the bottom left corner to 'Low', shifting its position in the chart one cell up, as in figure. Step 9 cannot be ignored (as a "Don't worry" category) any more.

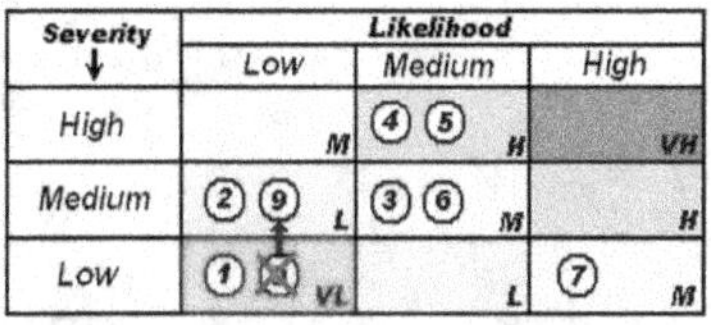

Fig. 15.3. - Change in risk status

However, this is not a critical situation. Attempts may be made to bring the equipment to original 'Low' severity level, or replace it. At worst, it may deserve closer monitoring.

On the other hand, a risk category may also improve, as can happen with the personnel becoming more used to the routine, with improvements in maintenance and supervision, etc. in which case the status may be updated to a lower risk level.

With changes in risk status being tracked at regular intervals, review of the series may indicate trends in safety management and safety culture of the company, useful as a predictor or 'leading indicator' for future decisions.

Thus the risk matrix is a dynamic visual tool to monitor and manage the risks in a job in a variety of modes.

———

16. FOLLOW-UP

The risk assessor's job is not over with completing the risk assessment and recommending controls. The closure must include follow-up. Some of the relevant topics have been mentioned in Chapter 5. Here, earlier comments are elaborated, and additional related matters are discussed.

16.1. RESIDUAL RISK

One would presume (or at least hope) that after risk assessment and implementation of existing and additional risk controls, the remaining risk, referred to as 'residual risk', will be less than what the RA team started with.

It would be good to demonstrate this improvement. So, a very common practice is to add a second set of assessment columns (for 'Likelihood', 'Severity', and 'Risk' levels) grouped as 'Residual Risk', and display the revised values, after implementation of the additional controls.

Of course, before the project can start or proceed, any remaining 'High' risk tasks would have been (a) eliminated, (b) farmed out, or (c) brought down to a lower (tolerable) risk level by additional investment and effort. Some of the 'Medium' risks would have been reduced to 'Low'.

A visual check will usually suffice to satisfy the assessors and the management. The improvement can also be quantified by assigning numbers to the various qualitative risk categories (with 1 for the lowest, 2 for the next higher and so on) and comparing the totals before and after RA and RM.

For the numerical assessment, we need to simply add the risk indices before and after the controls.

16.2. REVIEW OF ASSESSMENT

In the beginning, it may well happen that a team all of whose members do not like the sight of blood or the sensation of pain, might take the attitude: *"We cannot have any injury in our workplace!"* and then all the initial risk assessments may end up 'High', with only emergency measures like ambulances, police and fire departments as additional controls.

Or an over-enthusiastic team may continually impose many additional controls and take the stand: *"Well, these things happen, you know; let us not worry about it – insurance will take care of it!"* Then all risks may become 'Low'.

Thus, when all initial risks are 'H' or 'L', the team has chosen extreme positions, not reflecting good practice.

If all residual risks are 'High', the company will probably not be able to bring all of them down to 'Medium', and may as well close up shop.

If all residual risks are 'Low', then the company will have no more worries! Nobody needs to be concerned about any potential danger, except occasionally to glance around and check if anything is wrong. Complacency would set in, until suddenly something unexpected and terrible happens, when nobody is ready for it, nobody is in charge of it.

To avoid both extremes, the assessment team should review the entire spectrum of risks for the particular job or project, and try to re-sort them into three groups: 'Acceptable', 'Unacceptable', and 'Tolerable'.

However, by the end of the exercise, it will be necessary for residual risks to end up with only 'Medium' and 'Low'. That will confirm that the controls have been able to bring any 'High's down to 'M' or 'L'.

For a company with good safety culture, the resources and dedication to do a good job with safety, it will be quite reasonable to expect that after a few years, in jobs of the same kind they have being doing, the risk assessment for a fresh year will continue to remain the same as a previous year, except possibly for improvements in technological process or PPE.

Normally, if in one year an additional control has been recommended, it is proper to include it as ' existing control' for the succeeding year, as the required practice for the company if not the authority. Some of course may continue to list it as additional control year after year, to impress higher ups or regulators, which may also attract criticism.

Even if nothing untoward happens at the workplace, risk assessments are required to be reviewed at regular intervals (like 3 years in Singapore) and after every accident or serious incident, every major change in material, product, process, personnel or management, and necessarily after every regulation modification or introduction.

Regardless of requirements, it may be wiser to carry it out as an annual exercise, to serve as an evaluation point for annual budget. Higher risk jobs such as tunneling may rate even a quicker review. Singapore recommends a monthly review to catch problems before they become crises.

The RA team cannot rest on its laurels – or mope about its failures. It must be on constant alert reviewing the performance of the various safeguards and making sure that they continue to match up with the original – or latest – goals and standards.

The RA team must be constantly on its toes, as long as the company operates.

As the author sees it, a risk assessor's relationship to his duty is like an old-fashioned marriage: *"To have and to hold from this day forward, for better or for worse, for richer or for poorer, for better or for worse, in sickness and in health, till* (no, not 'death'!) *retirement do us part."*

16.3. RISK MATRIX FORMAT

As has already been discussed, the RA team and management may prefer to use the qualitative risk matrix format to start with, but it would be advisable to adopt the numerical format for use by field staff.

Likewise, the 3×3 is a good starting point. If extra rows or columns are added for distinct additional levels of severity or likelihood, or if extra categories are added to risk control, their definitions should be documented and disseminated to all concerned.

While the ('horizontal') format of the RA form presented herein is common and convenient, 'vertical' formats, or text type reports are also acceptable, if more suited to the industry, project, or a specific job.

In Singapore, trade-based risk assessments, such as for routine tasks with single well-defined input, process and output (E.g. plastering), are required only to submit a list of activities and (existing) controls.

The justification for this simplification would be that there would be very little deviation from the norm, and everybody involved would know and be experienced in their individual responsibilities and activities.

Every page of every form <u>must</u> carry expansions of abbreviations, interpretation of notations, and risk matrices for decisions, so that the user does not have to guess from memory, or refer to the first page or to the manual – which may tempt new staff to just repeat earlier entries!

In today's computerized world, this repetition does not take any more time than doing it the first time.

Consolidating all the features discussed up to this point, a typical risk matrix form, very similar to what Singapore uses, is presented as Table 16.1.

16.4. IMPLEMENTATION

It would be wise for the RA team to follow through by monitoring the implementation of the additional controls.

Table 16.1. TYPICAL RISK ASSESSMENT FORM

Department:		RA Leader (Name/Designation, Signature/Date):		Reference No:		Risk Matrix Size (LKL×SEV):
Process:				Risk Matrix (Circle): Qualitative/Numerical/Quantitative		
Process/Activity Location:						
Original Assessment date:		Approved by (Name/Designation, Signature/Date):				
Last review date:						
Next review date:						

Risk Matrix:

Severity ↓	Likelihood		
	Low	Medium	High

IMAGE OF RISK MATRIX

Item No.	HAZARD IDENTIFICATION			RISK EVALUATION					ADDITIONAL RISK Controls	RISK CONTROL					
	Work Activity	Hazard	Consequences	Existing Risk Controls	LKL	SEV	RISK		Additional RISK Controls	LKL	SEV	RISK	Implemen-tation Person	Due Date	Remarks
1															
2															
3															
4															
5															
6															
7															
8															

As mentioned in the previous chapter, whoever fills up the Risk Assessment form <u>must</u> (at least tentatively) enter the name and/or designation of an official assigned to implement and report on each specific additional control, and a date by which such implementation and reporting must be completed.

It is all too easy for the assessor to argue that right when he is filling up the RA form he would not know exactly who should implement the control or precisely when, and so postpone the assignments for later. This would be unwise and could be counter-productive.

It would be better to assign it to some (qualified) official and put some (reasonable) date before sending it to the concerned department, than to leave blanks. The assigned official must react; the clock will start ticking. If the initial assignments were not appropriate, changes may be made as and when necessary.

Without such a staff assignment nobody would do it, and without a date assignment the implementation process would simply not start.

A feedback and reporting mechanism should be in place to monitor progress of implementation, and manage any problems that may arise.

Once the risk assessment has been completed, additional controls have been recommended, and assignments have been made, all the concerned stakeholders must be informed of the findings and decisions, and their significance, at appropriate levels and with appropriate detail.

16.5. RECORD KEEPING

Maintaining records for three years will serve:

(a) As documentation for follow-up of long-term slow-acting hazards, and,

(b) As evidence of positive actions taken to eliminate or mitigate risks, in case an accident happens and the company has to prove its diligence in the matter.

The risk register is by now a thick folder, or even a series of files, containing hardcopies (text, graphics, photographs, spreadsheets) of every single thing that the team and management discussed, decided, and did up to that point.

Some may feel that the documentation of any mistake should be corrected in or removed from the register. The author reiterates that it may be better to leave everything in, making sure only that the mistake has been rectified promptly and well.

Not only wouldn't anyone in authority believe that any company would not have made even a single mistake, but also might an official take it as evidence of incomplete or inaccurate records – leading to the question: *"Who knows what other information may be missing?"*

What authorities would really want is solid evidence of planning and execution of risk assessment and control:

(a) from scratch (not photo-copied and altered from another industry, company, or project),

(b) on a continuing basis, and

(c) in a pro-active and participatory fashion.

Actually, full and detailed documentation will have many other side benefits. For one thing, writing down

Safe Work Procedures (SWPs) will be quite straight-forward now, because the details of hazards and controls can simply be expanded to step-by-step descriptions of how to carry out the hazardous tasks while implementing safeguards.

16.6. SAFETY AUDIT

An effective way to keep track of the completeness and effectiveness of all risk management activities would be to have audits.

Audits may be internal, by company staff, usually from another department, or external, by outside specialist agents.

An internal audit is much less expensive than an external one because the latter involves a fee, often hefty. But internal audit is beset with the predictable handicaps such as:

(a) pressure to finish the audit quickly without making too many waves;

(b) failing to recognize hazards or absence of controls due to over-familiarity of terrain, processes and people;

(c) hesitation to document safety deficiencies or violations of a fellow professional in the same company, preferring oral suggestions instead; and,

(d) conversely to (c) tendency to exaggerate shortcomings to make an example or return a grudge.

On the other hand, external audit will be often be

worth the fee, because:

(a) the company is paying them to find fault;

(b) the auditors are trained for the specific job they are employed to audit and if they miss to identify any hazard or fail to document it, their reputation will suffer, and in many cases, they can be sued too; and,

(c) it is better that the auditors find the deficiencies and violations and the company fixes them rather than the regulators find them and heavy fines or (worse!) stop work orders are issued.

16.7. EMERGENCY PREPAREDNESS

While all this planning and implementation of elimination or mitigation of risks are going on, let it not be forgotten that there is an all-enveloping criterion that must be addressed whenever we manage risks:

What to do in emergencies, when the safeguards fail, or in spite of their presence hazards escalate, or when new hazards appear unexpectedly? The response to this question is: 'Emergency preparedness'.

For instance, natural disasters are usually national emergencies, requiring a lot of planning and massive resources at the government level for crisis management, such as medical, traffic, environmental and other repercussions of the disaster.

Companies can only (and must) provide supportive equipment and services as needed or required.

In a workplace, emergency preparedness would take

following forms:

- First-aid kits, and for large enterprises, on-site first-aid clinics
- Fire extinguishers and fire-fighting equipment
- Detoxification chambers, safety showers, eye-wash fountains
- Rescue equipment, personnel and training
- Workplace emergency drills.

In many Asian countries, there is a cultural bias against talking about or sharing information on personal or professional failures or emergencies, and as a corollary, not much attention is paid to emergency procedures.

Author has found that – in a classroom of Asian engineers – where a fire extinguisher would be clearly visible, rarely would a person volunteer to pick it up and try to put out even a small indication of a fire within the room, preferring to run out and at best call the fire department.

[Author admits to himself having changed to the point he would use a fire extinguisher if and when necessary, because of his American stay!]

16.8. RESCUE MEASURES

While many companies know and implement risk control measures, not all of them are prepared for the subsequent rescue of the victims immediately after the accident from further injury or death.

Two examples will illustrate such rescue needs:

(a) Confined space rescue:

The dangers of oxygen deficiency and toxic gases in confined spaces are well known. Personnel are usually well prepared with oxygen and gas testers, gas masks, etc.

But what should be done when a worker inside a sewer faints due to some unexpected gas leak or some other trigger event?

A recent statistic states that over 60% of confined space fatalities are would-be rescuers. Before workers enter a confined space, they must have received proper training, and reviewed the SWP.

Rescue equipment comprising lifting tripods and pulley blocks, gas masks for rescuers, etc. must be ready at the worksite, and not have to be mobilized and brought to the site <u>after</u> the accident. *See* Fig. 16.1.

Fig. 16.1. - Rescue from confined space

Heroics of one or more outside personnel jumping in

to save their co-workers who have already been affected by toxic gas would be doomed to failure and lead to increased fatalities of the rescuers.

(b) Fall from height:

Almost everybody knows that the most common safeguard after falling from height is not a waist belt and cable, but a full body harness tied by a lanyard to an anchor, subject of course to co-requisites such as sufficient fall clearance and adequate anchors.

But many may not know that when a worker is thus "saved" from crashing to the ground and is hanging at the end of his lanyard and cable as in Fig. 16.2, he may have only about half an hour before he sinks into a coma and dies, due to what is known as 'venous pooling', contributing to 'suspension trauma'.

Fig. 16.2. - Rescue
of fallen victim

Not going into the medical reasons for it, let it suffice to say that the heart will be unable to pump blood up from the freely hanging legs after some time. (*See* author's website for more details.)

So, the fallen victim must be extracted from the suspended position and restored to the ground or other base as quickly as possible.

In Western countries which are used to a different culture of self-preservation from the Asians', self-rescue is part of a worker's training. This may involve his placing his feet into stirrups attached to his waist-belt or dropped to him from above, to restore circulation until help arrives.

Alternatively, he may release himself from his suspension ring and climb on the suspension rope, sometimes with foot-hold loops. Many pulley systems are available to facilitate self-rescue.

When the victim climbs back on a cable or rope ladder dropped to him, it is called 'assisted rescue'.

If the victim is not capable of self-rescue, either for lack of training or because he hit something on the way and is unconscious, then one or two rescuers will have to reach the victim from above by rope access, or from below on a boom hoist to extract the victim before he slips into suspension trauma, as shown in Fig. 16.2.

In any of these scenarios, it is critical that all procedures, equipment and trained personnel should be available at the site within minutes of the fall accident to be effective.

Such further rescue efforts must promptly be activated to bring him back to safety.

———

17. RISK COMMUNICATION

17.1. LANGUAGE OF COMMUNICATION

Even in countries like U.K. where the dominant language is the sole medium of communication, miscommunication happens due to the differences in the way the common language is used at different levels of the organization, the worker often failing to understand or appreciate the significance of what is being said or done.

Not only the language but also the idiom, syntax, and usage of communication within a single language become important.

Not surprisingly, the language we use becomes critical when many levels of management and execution of a project are involved.

Accidents are known to have been triggered, or made worse by lack of or poor communication between the workforce and line management.

(a) Problems with buzz words:

When technology is to be used by people with widely differing levels of verbal skills, the risk assessor should strive for the simplest of terms.

In risk management, terms like 'Remote' (likelihood), 'Catastrophic' (severity), 'Trivial' (risk) etc. may have to be defined in detail for users to memorize, if we wish to avoid ambiguity and confusion.

Graduates and well-trained staff may understand the

terminology, but think of the skilled and enthusiastic junior engineer or supervisor from another country, confronted with this jungle of jargon!

Author humbly (but with more than half a century of teaching experience!) suggests that companies stick to the simplest terms required for every occasion, if forms are to be distributed to and used by many levels of personnel, including the under-educated and insufficiently informed.

'Low', 'Medium', and 'High', abbreviated to 'L', 'M', and 'H', are good enough for three levels of safety-related measures.

Add 'Very Low' and 'Very High' at the ends, and you get five levels.

Maybe you may not have to go higher than five levels! If you must, 'Extremely Low', 'Extremely High', and so on may come in handy.

Likewise, avoid ambiguity in abbreviations. Not everybody can or will stop and place every abbreviation into the right context.

If there are two 'L's in a RA form, will 'L' mean 'Likelihood' or 'Low' when your foreman is in a hurry to fill it up? (This point was brought up in Chapter 9.)

(b) Flammable or Inflammable?

As a case in a point, there is confusion between 'Flammable' and 'Inflammable' on trucks transporting petrol, diesel etc. in different countries, and sometimes in the same country.

The original word in English for something that can

easily catch fire was 'Inflammable' rooted in the word 'Inflame'.

But over the years, when English spread around the world, new learners got used to 'In-' as a prefix to denote the opposite, such as 'Active'–'Inactive', and 'Justice'–'Injustice', to just cite two.

Gradually the original meaning of 'Inflammable' was blurred or lost, and the 'In' was dropped, with 'Flammable' preferred to denote fire hazard.

The National Fire Protection Association urged Americans in the 1920s to start using the word 'Flammable' to avoid confusion by people mistaking 'Inflammable' as meaning not being able to burn, and hence being careless around such signs and causing a fire – or even try to douse a fire with an 'inflammable' foam or liquid!

17.2. IMPACT OF DIFFERENT CULTURES ON SAFETY

Verbal and written communication becomes intertwined with shouts, gestures, body language, and other expressions of home culture.

The problem gets worse when more than one language becomes unavoidable in the workplace.

In recent decades, immigrant workers from South America flooded the southern parts of the USA – the influx still continues – and construction accident rates went up.

They discovered that it was often due to lack of communication between the locals and immigrants. Most signs, brochures, and training material, even websites were delivered in both English and Spanish. Foremen, supervisors and site engineers in these states were encouraged and even required to pass Spanish as a second language!

Needless to say, the accident rate for the immigrants dropped significantly.

17.3. THE SINGAPORE EXPERIENCE WITH COMMUNICATION

In Singapore, with its four national languages (English, Mandarin, Malay, and Tamil – the language of the earliest Indian settlers) and a dozen or so more languages from surrounding countries, has a special set of problems with communication with the immigrant laborers.

On top of it, the immigrants from a number of surrounding countries bring with them as many sub-cultures.

Considerable investment is being made to alleviate the problems in this area.

Such multi-lingual and multi-cultural mix of personnel compound communication problems with and among the immigrant labor. Personnel with different languages and from different cultures handle important phases of their projects.

Referring to 'Flammable' and 'Inflammable' again, the

confusion which has been overcome in the West still persists in the Eastern hemisphere, with many regions using the two words interchangeably, with not all viewers understanding their equivalence.

Although the British moved out of Asia half a century ago, tradition dies hard, and the Malay language still retains the word 'Inflammable' to denote fire potential.

Author has seen petrol-diesel trucks on the streets of a number of Asian countries with either word painted boldly on them!

17.4. GENERAL GUIDELINES FOR COMMUNICATION

- All communications must be in written form, with transmittal and receipt documented.

- Confirm your oral instructions in writing, as simply and unambiguously as possible, as promptly as possible.

- Encourage both your superiors and subordinates to seek clarifications – even disagree – instead of blindly accepting whatever you put down. It may be more "make-work", but it will be much safer.

- Don't hesitate to use an interpreter when addressing workers or even foremen on important issues. Ask if they have understood your spoken or written words. They may not volunteer information by themselves.

- Communication may be by memos and e-mail messages, posters and newsletters.

- Simple, clear, concise, and complete communication is a must in record-keeping and reporting – literally a matter of life and death.

For the first few times, and for significant new jobs, it may be worthwhile having a launch function for all the staff, to be followed by smaller group meetings of key people for interaction and feedback, regularly communicating up and down and across the chain of command, so that everybody becomes used to keeping up with – and even looking for – progress reports and newsletters on the effort.

Participation by all the stakeholders and all employees must be encouraged. Comments and suggestions should be welcomed.

To raise the staff morale, follow-up on the comments and suggestions must be reported regularly, even if it is an admission of delays or new problems.

Good safety work and effective safety related suggestions should be recognized and rewarded as appropriate. Constructive criticism and even negative feedback if correct, should be accepted with grace.

Senior management should take the lead and serve as role models in all the pro-active and participatory activities.

The philosophy of reporting violations and incidents, and practice of the new safety culture, must be nurtured by leaders adopting them, and by citing examples from company projects, newspaper reports, etc.

Risk communication also includes transference of information and knowledge, as well as on the spot guidance to personnel across the entire chain of command up and down from the CEO to the worker, and horizontally across departments.

Risk assessors must in fact be alert to communication problems among and between field staff which have appeared in recent past or which can be anticipated and make extra efforts to monitor and manage sensitive and dangerous areas in this regard.

———

18. WORKED EXAMPLES

18.1. EXAMPLE 1: RECEIVING AND STACKING GOODS IN SUPERMARKET

The job is to receive deliveries and move goods to their proper stacks in the storeroom of a supermarket. Maximum weight of a single package is 40 kg. Contents include corrosive cleaning agents hazardous to the eye and skin, and flammable liquids. Storeroom has mobile step-ladder. One checker and one worker are available. Assessed by qualitative method.

(a) Notes on risk assessment:

A worker should not be made to or allowed to carry more than 25 kg on regular basis.

Safeguards such as the eyewash fountain (for chemicals) and fire extinguisher must be checked for proper functioning at regular intervals.

Table 18.1 shows the risk assessment form for Example 1.

(b) Commentary on analysis:

1. (Hazard 2.1.) Manual Handling (Note 1): Extra worker to carry load will halve the load and thus reduce severity, shifting severity from H to M. But it will not change the frequency of the hazard, which thus will remain M as before. The risk reduces from 'H' to 'M'.

2. (Hazard 2.3.) Chemical spills (Note 2): Goggles will not reduce the frequency of the splash, but will

certainly prevent chemicals from hitting the eye and thus reduce the severity from 'H' to 'L', and thus reducing the likelihood from 'M' to 'L'. Risk reduces from 'H' to 'M'.

Table 18.1.- Example 1 : Delivery and stacking of goods at supermarket

Job Title: Delivery and stacking of goods | Assessor: John Doe | Date: 24/2/2007

No.	Activity	Hazard	Consequence	Existing or required controls	Risk Assmt L^*	S	R	Additional controls	Staff	Date	Residual Risk L^*	S	R
		Hazard Identification						Risk Control					
1.	Checking deliveries	Missing or wrong items, poor quality	Loss, Customer dissatisfaction	Checker	L	H	M	(None)	-	-	L	H	M
2.	Unloading goods, moving to store room, unpacking cartons, and stacking	2.1. Manual handling	Back pain	Trolley	M	H	H	Extra worker	Abc	+3WD	M	M¹	M
		2.2. Dropping cartons or goods	Loss, hand/foot injury, delay	Work gloves, shoes, training	M	L	L	(None)	-	-	M	L	L
		2.3. Spillage of toxic chemicals	Eye injury, skin injury	Eye-Wash fountain, first-aid kit	M	H	H	Goggles	Def	+2WD	M	L²	M
		2.4. Fire	Injury, death, property damage	Fire extinguisher	M	H	H	Fire drill	Ghi	+3WD	M	M	M
3.	Unpacking cartons	Cutter tool injury	Hand injury	Gloves	M	M	M	Training	Jkl	+2WD	L	M	L
4.	Stacking items on racks	Fall by wrong use of ladder	Major injury (death), damage to goods	Training	M	H	H	Extra worker	Mno	+3WD	L³	H	M

Date notes: +nWD = Actual date, n working days after RA form distribution date, to be entered.

Residual risk notes: 1 - Extra worker reduces load to half; 2 - Goggles prevent eye hit, 3 - Extra worker holds ladder.

→ Risk Matrix
L^* = Likelihood, S = Severity,
R = Risk, L = Low,
M = Medium, H = High

S↓ L*→	L	M	H
H	M	H	H
M	L	M	H
L	L	L	M

3. (Hazard 2.4.) Fire: Because of flammable liquids, likelihood of fire was taken as 'M'. A fire drill at regular intervals may not affect the frequency of fire, but will serve as a reminder to use the fire extinguisher to reduce the severity of consequences from 'H' to 'M', and hence the risk from 'H' to 'M'.

4. (Item 3.) Hand injury: Just like goggles, gloves may not reduce the frequency of the mishap. With additional training (and monitoring), the worker will take more care, reducing the likelihood from 'Medium' to 'Low'. But if there is a mishap, the severity will still be 'M'. Thus, M-M-M becomes L-M-L, one category of risk less.

The anomaly between 3. and 4. above is that while gloves have been listed under existing control, goggles have been listed as additional control. If the implication is that this was the first time that eye hazards were identified, then at least from the next review, goggles should be included as existing (required) control.

5. (Item 4.) Rack stacking (Note 3): By holding the ladder and monitoring and guiding the worker on the ladder, the likelihood of the fall is reduced from 'M' to 'L', but if the worker falls, the severity will still be high. So M-H-H reduces to L-H-M.

(c) Result of risk assessment:

The main result is that all 4 'High' risks have been brought down to 'Medium', enabling work to proceed. One 'Medium' has been brought to 'Low'.

With L = 1, M = 2, and H = 3 for risk levels, before controls, the total risk was 17, and after controls it is 12.

2

2

This improvement of 5/17 or 29% is only a nominal measure, not to be taken literally.

18.2. EXAMPLE 2: CUTTING GRASS WITH AN ELECTRIC LAWN MOWER

A person cutting grass with an electric lawn mower is exposed to many hazards. This is analyzed by the numerical method.

(a) Notes on risk assessment:

This is a general, 'non-technical' activity to illustrate the concepts of risk assessment. As most Westerners mow their own lawns, they would be quite familiar with the hazards, and from example and by experience learnt to avoid most of them.

Author too, having done it himself while in the USA, uses this as a 'common sense' case study from the home-owner's point of view.

When organizations (and home-owners) employ others to do the lawn mowing, they may have to show a formal risk assessment and control document.

Table 18.2 shows the risk assessment form for Example 2.

(b) Commentary on analysis:

1. 'Existing controls' are from experience, and apply more to protect hands and feet from injury and avoid slips and falls, than to take care of long-term and personal consequences, like hearing and breathing problems.

Table 18.2. - Example 2 : Cutting grass with electric lawn mower
(L = Likelihood, S = Severity, R = Risk, Res. = Residual, W-W = Who and When)

No.	Activity	Hazard Identification		Risk Assessment				Additional controls	Res. R			W-W
		Hazard	Consequence	Existing control	L	S	R		L	S	R	
1.	Cutting grass	1.1. Damaged cable	Electrocution	Cables / equipment approved standard	2	3	6	ELCB / RCD, Inspect before & during use	1	3	3	*
		1.2. Noise	Hearing loss	(None)	3	2	6	Ear plugs	3	1	3	*
		1.3. Wet grass	Slip, fall	Rubber soled shoes	2	2	4	Extra care, no cutting in rain	1	2	2	*
		1.4. Rotating blade	Foot injury	Regular shoes	2	3	6	Steel-toed shoes	2	1	2	*
		1.5. Trailing cable	Cut cable, electrocution	None	3	3	9	Spring retractor, ELCB	1	3	3	*
		1.6. Slippery surface	Slip, fall	Regular shoes	2	2	4	High-grip soled shoes	1	2	2	*
		1.7. Debris in grass	Limb, face, or eye injury	Work gloves and shoes	3	2	6	Goggles, full-body clothing	3	1	3	*
		1.8. Cut grass / pollen in air	Allergy, asthma	(None)	3	2	6	Preventive medication, attach bag for cut grass	2	1	2	*
2.	Lifting mower or cut debris	Excessive load, wrong posture	Back pain, muscle pull	(None)	3	2	6	Training, helper to move mower, wheeled cart for cut grass	2	1	2	*
				Total			53	Total			22	

* - Mostly done by mower, before and during mowing

[Note: The accompanying picture is not intended to be a recommendation for grass mowing. Many hazards are present, to be eliminated or managed]

Risk Matrix

Severity	Likelihood		
	1 (Low)	2 (Med)	3 (High)
3 (High)	3	6	9
2 (Medium)	2	4	6
1 (Low)	1	2	3

Risk Categorisation

Risk Category	Risk Index Ranges	Risk Designation
I	1-2	Low
II	3-4	Medium
III	6-9	High

2. There are no mandatory regulations as such for managing lawn-mowing hazards. Most westerners grow up with technological awareness and confidence.

183

3. Items 1.2., and 1.5: The ELCB cuts down the likelihood of electrocution to very low, and the spring retractor reduces chances of cable damage.

4. Items 1.5, 1.8., and 2: Existing controls '(None)' simply reflects the fact that people generally do not identify hazards until it happens to them, or until accumulated effects force them to analyze the causes.

   So in Item 1.7 for instance, nobody would wear goggles for lawn-mowing until some debris hits them in the eye, and even then, they may just decide to be more careful rather than start wearing safety goggles!

5. (Hazard 1.8.) Allergy, asthma: Preventive medication will not reduce frequency of exposure but will reduce severity of asthma attack. Bag attachment to collect cut grass will definitely reduce the amount of pollen and dust reaching the victim's nose, reducing the likelihood. Thus 3-2-6 becomes 2-1-2.

6. (Item 2.) Helper and wheeled mechanical aid will reduce likelihood of lifting heavy loads. Training for safe lifting will reduce the severity of the stress on backbone. Here also, 3-2-6 reduces t o 2-1-2.

(c) Result of risk assessment:

All the 6s and 9s of risks have been brought down to 1s and 2s. Here the improvement is (53-22)/53 i.e. 31/53 or 58%, quite dramatic.

18.3. COMMENTS ON EXAMPLES 1 AND 2

1. Both examples are only for illustrative purposes. The

two formats are nearly same. Note abbreviations and risk matrix are included.

2. Everything from the format up to the assignment of the levels, right up to the controls, must reflect the company's understanding of its problems and its commitment to solve them.

3. Four 'M' risks and five '3' (= 'M') risks remain in examples 1 and 2 respectively. No further control from the hierarchy is feasible, and these items can and must be managed safely.

4. Compliance regime is assumed in all cases.

It may happen that not knowing or forgetting the underlying principles, or in haste or carelessness, the assessor may assume a control to reduce likelihood instead of severity or vice-versa, or even to reduce both likelihood and severity.

Author feels it would be necessary to occasionally remind oneself that the two key phrases for controls are:

(a) 'Likelihood of occurrence of the mishap' – not the likelihood of reduction of the consequence; and

(b) 'Severity of consequences of the mishap', if and when it occurs.

Even so, philosophically (or practically?) speaking, which of the two factors is reduced by a control may not matter as much as whether the positive effect of a control has been included in one or the other of the two factors.

The only caution to be remembered here is that many get into the habit of reducing both likelihood and severity with the introduction of a control, without examining the justification of such a double benefit – as has been shown in Items (B) 5 and 6 of Section 17.2 above! This should be avoided at all costs.

18.4. EXAMPLE 3: GOODS TRANSPORTATION (LOGISTICS) ANALYIS

We shall carry out the risk assessment of a goods storage and transportation ('Logistics') company for the moving and delivery job.

Figure 18.1 depicts certain common activities at the company for which risk assessment has to be carried out from loading at the warehouse up to delivery to the customer.

Fig. 18.1. - Activities at a warehouse

Goods include flammable materials also.

Work-related transportation safety is to be included

in the analysis.

We will use the 3×3 qualitative risk matrix with three levels of likelihood and three levels of severity, the resulting risks to be grouped into three risk categories, as in Fig. 11.4.

We will further assume that only statistical data on frequency of occurrence and severity of resulting physical harm and property damage is available.

The example is intended to be illustrative of general principles and may not conform to any particular risk management regulation.

Only five activities and their corresponding hazards and consequences in the personal harm and property damage domains will be considered.

There are sure to be other activities equally or more critical to be analyzed in an actual situation.

Step 1: *Listing of activities, hazards, consequences and existing controls:*

In Table 18.3 listing all the above-mentioned quantities for the five activities as numbered, note the following:

1. <u>Driver fatigue:</u> due to excessive overtime is a frequent hazard on long-distance runs.

   A resulting accident will cause (a1) major injuries or death, and (a2) some property damage.

Table 18.3. Example 3 - Transportation RA, Hazard, consequences
and existing controls

No.	Activity	Hazard	Consequences	Existing Controls
1.	Driving	(a) Driver fatigue	(a1) Major accident, Major injury, fatality	Proper schedule-ing
			(a2) Major accident, Moderate property damage to truck and goods	- Do -
2.	Prepara-tions for trans-port	(a) Load securing, unloading, Coupling and uncoupling trailers	(a1) Crush injuries to hands and feet	Training, gloves
			(a2) Minor property damage	- Do -
3.	Moving flam-mable goods	(a) Fire	(a1) Burns	Fire Extin-guishers, fire drills
			(a2) Major fire, Major property damage	- Do -
4.	Collecting and deliver-ing mate-rials	(a) Slips and trips at depot	(a) Cuts and bruises	Rubber mats, and non-slip shoes
		(b) Slips and trips at customer sites	(b1) Cuts, bruises, and fractures	Non-slip shoes
			(b2) Minor property damage	- Do -
5.	Handling goods	(a) Manual handling at depot	(a) MSD, back pain, cuts, bruises	Forklift Limit carry wt. to 25 kg, training
		(b) Manual handling at customer site	(b) MSD, back pain, cuts, bruises	- Do -

2. <u>Preparations for transport:</u> Load securing and unloading, and coupling and uncoupling of trailers can cause (a1) injuries to hands and feet, as well as (a2) damage to dropped goods.

3. <u>Moving flammable goods:</u> This may occasionally have a fire hazard. That may cause (a1) burns to the body, but even if all the people have been evacuated, the conflagration can result in (a2) a lot of property damage.

4. <u>Collecting and delivering of materials:</u> This involves the personnel carrying or carting parcels around the floor with slip and trip hazards.

   Workers will have non-slip shoes, but while at the home depot, the floor may have non-slip rubber mats, the same may not be expected at the customer sites.

   Hence, the hazard must be examined separately:

   (a) at the depot, and

   (b) at the customer site, with a greater risk of (b1) cuts and injuries and (b2) minor property damage at the latter.

5. <u>Handling Goods:</u> Throughout the operation, workers will be exposed to manual handling of goods, hopefully restricted to 25 kg per person, expected to be trained in lifting and carrying the packages.

   Even so, repeated carrying with improper posture will pose a MSD hazard. This hazard (a) may be well controlled at the home depot, but (b) may be open to greater risk at the customer site.

   These considerations have been tabulated in Table 18.3 along with existing controls.

To make this example quite general, author will assume that no detailed guidelines are available to assess the likelihood and severity levels.

Hence we will not be following the three-level descriptions accompanying the 3×3 risk matrix shown in Tables 9.1 and 10.1.

We will instead adopt the author's 'kick-start' method of listing all likelihoods and severities and then grouping them into three levels, as suggested in Sections 9.6 and 10.7.

Step 2: Listing all likelihoods in the job

Table 18.4 lists the likelihood data for the various known industry-wide hazards or estimated by the RA team.

Table 18.4. Example 3 - Transportation RA, Known/estimated likelihoods

No.	Hazard	Likelihood (w. Controls)
1.	(a) Driver fatigue	Once a month
2.	(a) Load securing, unloading, etc.	Once a week
3.	(a) Fire from flammable goods	Once in 10 years
4.	(a) Slips and trips at depot	Once a year
	(b) -Do- at customer site	Once a month
5.	(a) Manual handling at Depot	Once a year
	(b) Manual handling at customer site	Once a week

Note that in items 4 and 5, for the same hazards the frequency of mishaps are different for the depot and the customer site.

Step 3: Defining the likelihood level criteria

We now note that the likelihoods vary from a low of

once in 10 years to a high of once a week.

So we decide the likelihood level criteria as shown in Table 18.5, remembering to include a larger portion of the range in the 'Medium' segment, subject also to the condition that the covered frequencies are 'acceptable', 'tolerable', and 'unacceptable' in the 'Low', 'Medium' and 'High' respectively.

Table 18.5. Example 3 - Transportation RA, Likelihood level criteria

LKL Item	Low	Medium	High
Frequency	> 2 years	2 years to > 1 week	<= 1 week

Step 4: Sorting the items according to the set criteria

Table 18.6 shows the likelihoods from Table 18.4 arranged in increasing order, with the corresponding items grouped together, and their respective likelihood levels.

Table 18.6. Example 3 - Transportation RA, Likelihood assessment

No.	Mishap Frequency	Items	Likelihood Level
L1	Once in 10 years	3(a)	L
L2	Once a year	4(a), 5(a)	M
L3	Once a month	1(a), 4(b)	M
L4	Once a week	2(a), 5(b)	H

Step 5: Listing all severities in the job

Table 18.7 lists the severity data for the various known industry-wide hazards or estimated by the RA team.

Note that only item 5 has the same severity for both its components.

Table 18.7. Example 3 - Transportation RA, Known/estimated severities

No.	Consequences	Severity (w. Controls)
1.	(a1) Major accident injury, fatality	MC > 6 months
	(a2) Moderate property damage	$150,000
2.	(a1) Crush injuries to hands and feet	MC, 2 months
	(a2) Minor property damage	$20,000
3.	(a1) Burns	MC, 3 months
	(a2) Major property damage	$500,000
4.	(a) Cuts and bruises	MC, 3 days
	(b1) Cuts, bruises, and fractures	MC, 3 months
	(b2) Minor property damage	$10,000
5.	(a, b) MSD, back pain, cuts, bruises	MC, 2 weeks

Step 6: Defining the severity level criteria

We now note that the physical harm severities vary from a low of 3 days MC to a high of MC of more than 6 months (or fatality), and property damage severities vary from a low of $10,000 to a high of $500,000.

So we decide the severity level criteria as shown in Table 18.8, remembering to include a larger portion of the range in the 'Medium' segment, subject also to the condition that the covered severities are 'acceptable', 'tolerable', and 'unacceptable' in the 'Low', 'Medium' and 'High' respectively.

Table 18.8. Example 3 - Transportation RA, Sverity level criteria

SEV Item	Low	Medium	High
Injury, MC	<=1 week	>1 week to 6 months	> 6 months
Property, $	<= $10,000	$10,001 to $200,000	> $200,000

Step 7: Sorting the items according to the set criteria

Table 18.9 shows the severities from Table 18.7 arranged in increasing order, with the corresponding items grouped together, and their severity level.

Table 18.9. Example 3 - Transportation RA, Severity assessment

No.	Mishap Consequence	Items	Severity Level
	Personal Injury		
S1	MC, 3 days	4(a)	L
S2	MC, 2 weeks	5(a, b)	M
S3	MC, 2-3 months	2(a1), 3(a1),4(b1)	M
S4	MC, > 6 months (& fatal)	1(a1)	H
	Property Damage		
S5	$10,000	4(b2)	L
S6	$20,000 – $150,000	1(a2), 2(a2)	M
S7	$500,000	3(a2)	H

Step 8: Determining the risk and recommending controls

Major – critical – part of the risk assessment is over. Now the likelihood and severity levels recorded in Tables 18.4 and 18.7 and categorized in Tables 18.5 and 18.8 may be tabulated and the two factors combined by the 3×3 risk matrix of Fig. 11.4, reproduced here for convenience of reference.

Severity ↓	Likelihood		
	Low	Medium	High
High	Medium	High	High
Medium	Low	Medium	High
Low	Low	Low	Medium

Fig. 11.4.-3 by 3 risk matrix, 3 categories

The resulting risk values are shown in column 5 of Table 18.10. We see that there are 3 'L's, 4 'M's, and 4 'H's.

The 'High' risks must be eliminated or brought down to at least 'Medium'.

Table 18.10. Example 3 - Transportation RA, Risk assesssment and control

No.	Consequences	Original Risk			Additional Controls	Residual Risk		
		LKL	*SEV*	*RISK*		*LKL*	*SEV*	*RISK*
1.	(a1) Driver fatigue, major injury, fatality	M	H	H	Strict over-time control	L	H	M
	(a2) Driver fatigue, moderate property damage	M	M	M	- Do -	L	M	L
2.	(a1) Loading, crush injuries	H	M	H	Re-training, extra loading helpers, extra supervision	M	L	L
	(a2) Loading, minor property damage	H	M	H	- Do -	M	M	M
3.	(a1) Fire, burns	L	M	L	(None)	L	M	L
	(a2) Fire, property damage	L	H	M	(None)	L	H	M
4.	(a) Slips and trips at Depot, Cuts and bruises	M	L	L	(None)	M	L	L
	(b1) Slips and trips at customer site, fractures	M	M	M	(None)	M	M	M
	(b2) Slips and trips at customer site, property damage	M	L	L	(None)	M	L	L
5.	(a) Manual handling at depot, MSD, back pain	M	M	M	(None)	M	M	M
	(b) Manual handling at customer site, MSD, back pain	H	M	H	Send extra helper to site	M	L	L

<u>Items 1(a1) and 1(a2):</u> Driver fatigue is easy to

control, by increasing the monitoring of driver logs (and their health condition on the day of their assignment) and strictly controlling over-time.

This should reduce the likelihood from 'M' to 'L' – but will not reduce the severity if and when an accident occurs. Consequently risks for both the items will reduce from 'H' and 'M' to 'M' and 'L'.

Item 2(a1): The frequency of crushing injuries to hands and feet during loading operations seems excessive. A close review may reveal root or contributory causes which may be rectified.

Meanwhile, re-training, providing additional helpers and having extra supervision during the loading operations should reduce the likelihood of injury from 'H' to at least 'M', and (because to the sharing of the wok) possibly the severity of the injury also from 'M' to 'L', thus reducing the risk from 'H' to 'L'. Such a drastic shift would be rare, and worth monitoring for continued validity.

Item 2(a2): The frequency of property damage during loading operations tied to injuries, also seems excessive. Again a close review may reveal root or contributory causes which may be rectified.

Meanwhile, re-training, providing additional helpers and having extra supervision during the loading operations should reduce the likelihood of injury from 'H' to at least 'M', but the severity of the damage may not reduce if and when it happens.

Item 5(b): Manual handling of loads is always a problem, but obviously it could be better controlled at the home depot, leaving 'High' likelihood for customer site work.

If an extra helper is sent to customer site, both the likelihood and severity may be reduced one level (subject to confirmation at site), thus reducing the risk from 'H' to 'L'.

We have now eliminated the presence of all four 'High' risks, reducing them to 'M' or 'L'. These changes are marked in the Residual Risk columns in Table 18.10.

It just happens that all the Additional Controls happen to be Administrative. In this example none of the other four controls would be feasible.

To quantify the improvement – although they were necessary ones – we may use the method of Example 1, by which the original risk index sum was 23, modified to 17, an improvement of 30%.

All that is left – but not shown – is to assign a person (by name or designation) to carry out the recommended additional control, and to set a specific frame to complete the task. The job may now commence.

The preceding appears to be a long and time consuming procedure. Much of it because every entry and decision has been explained.

Not to worry! Like anything else, once a beginner has become familiar with the process and a certain type of job, subsequent risk assessments should be much faster and more elegant.

What the author wishes to emphasize is that a risk assessor can easily fall into the trap of thinking that two jobs that appear to be very similar or forming part of the same project, will lead to the same risk assessment.

Not true! Except for manufactured goods, two very similar jobs done in different locations and at different times are sure to have at least a few differences that require a fresh look of the second one to seek out the precise differences and assessing and controlling them.

Too many accidents happen by such over-confidence with risk management, even for – or especially for – veteran assessors.

Better to be safe than sorry!

———

19. OTHER METHODS OF RISK ANALYSIS

In previous chapters, only one method of risk analysis has been discussed, namely assessments of likelihood of occurrence and severity of consequence of a hazard, followed by combination of the two factors to determine the risk outcome by means of a risk matrix.

Although such risk analysis utilizing the risk matrix is by far the most common assessment tool, there are other methods of risk analysis (or 'mishap analysis', to include the investigative phase) that have been, and continue to be, in use in various parts of the world, for specific purposes and in certain industries.

These methods focus on identification of hazards that may cause mishaps and on consequences that may flow from the mishaps in various ways, as will be described in the following sections.

19.1. METHODS FOR MISHAP ANALYSIS

Many methods are available for risk assessment and related topics, apart from the risk matrix approach.

These methods are generally designed to identify the causes and/or assess the hazards in job activities before a mishap occurs. Some of them may analyse the consequences of a mishap after it occurs relating them to the failure of risk safeguards and controls, rather than to the hazards themselves.

Both types are aimed at preventing and managing future accidents. They can play an important role in risk management.

A typical list is as follows:

1. ALARP (As Low As Reasonably Practicable), as already discussed in Section 11.13.
2. Job Safety Analysis (JSA) or Job Hazard Analysis (JHA)
3. 'What if ...?' Analysis
4. HAZOP (Hazard Operability)
5. FMEA (Failure Modes and Effects Analysis)
6. FMECA (Failure Modes, Effects and Criticality analysis)
7. FTA (Fault Tree Analysis) *
8. ETA (Event Tree Analysis) *
9. Cause and Effect ('Fishbone' or Ishikawa) Analysis

Of the above nine methods, the two marked with an asterisk, namely FTA and ETA, may be extended to enable quantitative assessment of the risk, if numbers can be assigned to each of the component factors contributing to the risk.

Finally, a 'Bow-tie' or 'Butterfly' Diagram' (called thus because of the shape of the graphic format of the process) is used to integrate both the before and after segments of a mishap, to provide the risk manager a complete picture of the accident sequence.

Following Section numbers are keyed to above list.

19.2. JOB SAFETY OR JOB HAZARD ANALYSIS

These are very common generic terms used to refer to identification of hazards and development of controls for any task in an industry.

USA, UK, and many countries advocate JSA (or JHA) through the following three steps:

1. Break selected job down into a sequence of steps
2. Identify potential hazards for each step
3. Determine safety measures to overcome these hazards

Note nothing is said about how hazards are assessed and how controls are determined, these aspects being left to the individual industries and companies.

In this book, we have already adopted this process, with much material to guide the risk assessor determine risk level from likelihood and severity levels, and how to decide the type of control to manage risk.

In that sense, we have already covered JSA/JHA.

19.3. 'WHAT IF ...?' ANALYSIS

(a) Principle and Procedure:

'What If...?' analysis is a 'Common Sense' method, not too structured. It is largely experience based and generally achievable by a group of specialists brain-storming on a topic. Its main aim is to identify the hazards and their consequences

The method involves asking ourselves and other stakeholders what if some mishap occurred in the various components of a job, such as human, mechanical, administrative, etc. The answers will automatically highlight whatever hazards there are.

Actually, the process only serves to identify the hazards, and does not define their levels. One may use the list of answers for further assessment and control.

'What-If ...' questions can be formulated to reflect human errors, process deficiencies and equipment failures, etc. which may occur during normal production operations, erection or construction, maintenance activities, and corrective/de-commissioning situations.

Some general situations which the questions could address may be the following:

- Procedural errors, such as incompleteness, incorrectness, etc.
- Human errors, such as, lack of training, inattention, ignorance, etc.
- Equipment errors, such as unsuitability, break-down, poor maintenance, etc.
- Administrative errors, such as inadequate training,, maintenance, inspection, etc.
- Instrument errors, such as mis-calibration, poor maintenance, wrongly read, etc.
- Utility failures, such as with power, water supply, steam, gas, etc.
- External influences such as bad weather, vandalism, fire, etc.
- Adverse combination of simple deviations

(b) 'What If ...?' Example

Table 19.1 lists many 'What if ...' questions and some answers for risk assessment of a concreting job.

Table 19.1. Sample 'What if ...?' questions and answers

No.	Question: What if ...?	Answer
1.	Foundation for falsework is not firm or strong enough?	Formwork will sink during casting and fail.
2.	Screw jacks do not have enough travel for adjustment during casting?	Will cause delay in rectification or unevenness in the finished structure.
3.	Provision for leak rectification is not made?	Will cause delay or lead to stoppage of work.
4.	Watcher cannot check for sinking or leakage from outside casting area?	May lead to collapse.
5.	Objects heavier than 25 kg have to be moved by worker?	Will cause back-pain and other MSD.
6.	Concrete delivery hose gets blocked?	Can lead to hose swing injury and damage.
7.	Communication walkie-talkie fails during concreting?	Will cause delay or lead to over-loading, and thus to collapse.
8.	The delivery truck is delayed too long or cancelled?	No injury or damage but loss of work and time.
9.	There is too much delay between ready-mix truck arrival and casting?	Concrete quality will suffer and/or the concrete will become unusable
10.	Bracings against buckling and sway are inadequate?	Formwork structure may fail and collapse
11.	Workers' limbs are not protected against cement contact?	Will result in cement burns
12.	Workers' eyes and nose are not protected against cement dust?	May result in irreversible eye and lung damage

The list may not be complete. It must include any and all concerns raised by all the stakeholders to be really effective.

Many items may turn out to be non-injurious and non-damaging, with only loss of work and time being involved, but even they may need accounting for and some kind of pre-planning.

Author has included Item 7. on communication failure solely because that is precisely what happened during a concreting job when the author was observing:

- A supervisor was guiding the crane operator with the command *"Down/Down/..."*, to lower a hopper of concrete on to a formwork. His walkie-talkie suddenly failed but the hopper still continued to come down.

- Supervisor and crane operator could not see each other. The supervisor had to dash down two floors to catch the eye of the crane operator and get him to stop the hopper – a few centimetres above the formwork, in a *'Mission Impossible'* type climax!

- If he had been a few seconds late the heavily loaded hopper would have hit the formwork and crashed it to the ground, carrying three workers with it!

- What did I learn from it? That we should have back-up walkie-talkies on critical jobs!

Item 8. refers to a relatively minor hazard, but most of the others may lead to medium or major hazards.

This method is mainly intended to document potential problems and try to be prepared with solutions and controls in case they happen.

The method of Table 19.1 may be extended with

columns for 'Consequences', 'Likelihood', and 'Severity' – or equivalent considerations to assess the risk, after which further action can be taken to manage the risks by suitable action, including but not limited to the risk matrix method discussed earlier.

19.4. THE HAZOP METHOD

(a) Principle and Procedure:

'HAZOP' stands for 'Hazard Operability'. It is a systematic technique for identifying hazards and operability problems throughout an entire facility.

Instead of questions, the method presents a list of 'guidewords' and 'parameters' to identify deviations from norms during a process.

This is better than the *'What if ...?'* method for process hazard identification.

It is particularly useful to identify unwanted hazards which are:

a) Unknowingly designed into facilities due to lack of information, or

b) Introduced into existing facilities due to changes in process conditions or operating procedures.

Again subsequent risk assessment and control will follow conventional lines for specific industries.

The objective of HAZOP study is to detect any predictable deviation (undesirable event) in a process or system, by the systematic study of the operations in each phase.

The procedure may be summarised as follows:

- The system is divided into functional blocks.
- Every part of the process is examined for possible deviations from the design intention.
- Analysis must include each major item of equipment, and all supporting equipment, piping, instrumentation, and other accessories, for:

  a) Every phase of the process, and,

  b) Each system and operator.

- Any deviation from the norm that may credibly cause a hazard or inconvenience, is defined as the combination of a 'Guideword' and a 'Parameter'.
- Guideword is usually a condition listed as an adjective (E.g. 'Low').

  Parameter is generally a process factor listed as a noun (E.g. 'Pressure').

  The combination of the two factors (E.g. 'Low Pressure') will thus represent a deviation from the norm which could be a hazard in the process.

- Once all the deviations have been documented, each deviation is evaluated to decide how it could be caused and what the consequences would be. Subsequent risk assessments and controls will follow along the usual lines.

(b) Guidewords and Parameters:

Guideword adjectives are the common terms used to compare some feature of the process with the norm: E.g.: 'No' (meaning 'missing'), 'Low', 'High', 'Reverse' (or 'Opposite') etc., as appropriate to process and product.

Parameter nouns are the basic terms that refer to various activities and parts of the process or product: E.g.: 'Flow', 'Level', 'Pressure', 'Temperature', 'Agitation', etc., as appropriate.

Figure 19.1 shows some schematic combinations of guidewords and parameters. In the figure, some cells are left blank illustrating the situation where the combination may not be a valid part of the process.

GUIDE-WORDS →	NO	LOW	HIGH	PART OF	ALSO	OTHER THAN	RE-VERSE
FLOW	NO FLOW	LOW FLOW	HIGH FLOW	MISSING INGRE-DIENTS	IMPU-RITIES	WRONG MATL.	RE-VERSE FLOW
LEVEL	EMPTY	LOW LEVEL	HIGH LEVEL	LOW INTER-FACE	HIGH INTER-FACE	-	-
PRES-SURE	OPEN TO ATM.	LOW PRES.	HIGH PRES.	-	-	-	VACUUM
TEMPE-RATURE	FREE-ZING	LOW TEMP.	HIGH TEMP.	-	-	-	AUTO REFRIG.
AGITA-TION	NO AGI-TATION	POOR MIXING	EXCES-SIVE MIXING	IRREGU-LAR MIXING	FOAM-ING	-	PHASE SEPA-RATION
REAC-TION	NO RE-ACTION	SLOW REAC-TION	'RUN-AWAY REAC.'	PARTIAL REAC-TION	SIDE REAC-TION	WRONG REAC-TION	DECOM-POSI-TION
OTHER	UTILITY FAIL-URE	EXTER-NAL LEAK	EXTER. RUP-TURE	-	-	S1, S2, MAINTE-NANCE	-
PARA-↑ METERS	[REFRIG = REFRIGERATION, S1 = START-UP, S2 = SHUT-DOWN]						

Fig. 19.1. - HAZOP schematic

Once the hazard combinations are identified, they may be assessed to formulate appropriate controls.

(c) HAZOP Example:

Table 19.2 shows a simple sample HAZOP table for the delivery of liquid propane to a tank.

Table 19.2. HAZOP for delivery of liquid propane to tank

GUIDEWORD → PARAMETER ↓	MORE	LESS
LEVEL	More Level	Less Level
TEMPERATURE	More Temperature	-
PRESSURE	More Pressure	-

Of the six possible deviations (from the two guidewords and three parameters) 'Less Temperature' and 'Less Pressure' are either not feasible or not critical, and hence have been omitted in the Table.

Table 19.3 is an example of how the HAZOP table may be extended towards risk assessment and control.

19.5. FAILURE MODES AND EFFECTS ANALYSIS (FMEA)

(a) Principle and Procedure:

Objective of FMEA is to identify how failures could occur (failure modes and causes) and the consequences of failures on performance, equipment, personnel, etc., and their consequences on objectives (failure effects).

This was developed by the U.S. military during the 1940s and supported by military specification beginning in 1949. It was used by reliability engineers to determine problems that could arise from malfunctions of hardware.

FMEA is now firmly established as a risk analysis and risk management methodology. FMEA methods and applications are now recommended practice for aerospace engineering, certain processes and structures, and in healthcare.

Generally, following four failure modes are checked, although some special processes may involve more:

1. Premature operation of component;
2. Failure of component to operate at a prescribed time;
3. Failure of component to cease operation at a prescribed time; and
4. Failure of component during operation.

Corresponding failure causes, and their consequences are also analyzed, to extend the hazard identification to risk assessment and control.

(b) FMEA Example:

Table 19.4 displays part of NASA Spacecraft Flight Assurance Procedure specifically for a Wrist Actuator.

Table 19.4. FEMA for NASA's wrist actuator

Mission DTF - 1 System FTS Subsystem/Instrument 3.13 Component Wrist Actuator Mission Phase Orbit	**FLIGHT ASSURANCE PROCEDURE** *http://fmea-fmeca.com/FMEA_Nasa_spacecraft.pdf*	Date 8-10-96 Prepared by Ron Smith Approved by RHB

Failure Mode Number	Identification of Item or Function	a. Failure Mode b. Failure cause	Failure Effects a. Local or Subsystem b Next Higher Level - System c. End Effect - Mission	Severity Category	Remarks a. Failure Detection Method b. Compensating Features/Action c. Other
3.13.6	Wrist actuator, roll provides motion in roll (x) axis	a. Loss of motor control b. Part failure in motor drive circuit	a. Loss of wrist roll motion and torque b. Cannot continue FTS task and mission c. None at Orbiter mission	2R	a. Position sensor & torque sensor displayed at DAC b. Backup hardware to put arm in safe position. Good arm can put arm in safe position.

In the Table, Severity Category '2R' refers to the third (starting from 'Minor') of six categories NASA has defined for the topic. Reference may be made to the link provided in the Table. This kind of definition is an alternative to the determination of risk by the risk matrix.

Obviously, the FEMA method is quite similar to the qualitative method of risk analysis we have already discussed.

19.6. FAILURE MODES, EFFECTS, AND CRITICALITY ANALYSIS (FMECA)

(a) Principle and Procedure:

This is just an extension of FEMA to one higher level of 'Criticality' determination, by including a third factor reflecting ease or difficulty of detection of the problem.

Then 'Risk Priority Number' (RPN) is defined as the product of 'Severity' (S), 'Frequency' (F), and 'Probability of Detection' (D), i.e. **RPN = S*F*D**

As generally each of the three factors are assessed on a scale of 1 to 10, maximum RPN = 1000, higher the number reflecting greater risk.

Except for failure mode identification, this is same as numerical risk assessment method already discussed.

(b) FMECA Example:

Table 19.5 shows another example from NASA, this time of risk assessment for a tire.

Table 19.5. FMECA Example of a tire risk

Tire FMECA with Reevaluation of Risks

Part Name Potential Failure Modes	Causes (failure mechanism)	Effects	Risk Priority Rating				Recommended Corrective Action	Improved Rating			
			Sev"	Freq	Det	RPN		Sev"	Freq	Det	RPN
Cord Fiber separation	1. Weak precursor material	Ply failure	4	3	8	96	Incoming inspection	4	1	8	3
	2. Handling damage	Ply failure	4	3	8	96	Increase process controls during mfg	4	2	2	16
	3. Cumulative fatigue	Ply failure	4	2	8	64	Monitor tire life	4	2	2	16
Ply Delamination	1. Dirt or grease	Loss of side wall integrity	7	3	8	168	Toluene wipe down during layup	7	1	1	7
	2. Twisted plys	Loss of side wall integrity	7	2	6	84	Automatic ply alignment	7	1	1	7
	3. Poor bond pressure	Loss of side wall integrity	7	2	8	144	Redundant tensioning system	7	1	1	7
Carcass Disinte- gration	1. Poor tire alignment	Vehicle loss	9	2	9	162	Planned periodic maintenance	9	1	1	9
	2. Tire hits curb	Vehicle loss	9	2	9	162	Driver training	9	1	1	9

"Severity ratings 8 to 10 request special effort in design improvement regardless of RPN rating
http://www.hq.nasa.gov/office/codeq/risk/docs.incose.pdf

There is practically no difference between this and our Example 3 of the previous chapter, except for the number of factors contributing to risk (3 versus 2) and the number of levels of each factor (10 instead of 3).

19.7. FAULT TREE ANALYSIS (FTA)

(a) Causes and consequences of mishap:

Every mishap (accident or loss) is caused by a number of contributing causes and every mishap leads to a number of consequences, which may respectively be likened to the roots and branches of a tree.

This 'mishap tree' concept is illustrated in Fig. 19.2.

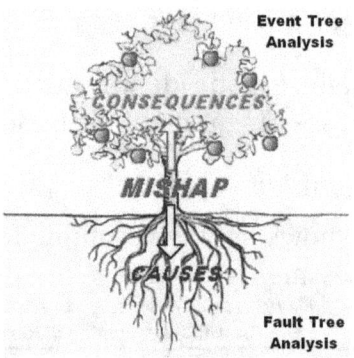

Fig.19.2. The Mishap Tree

The analyses for causes and consequences are respectively called 'Fault Tree Analysis' (FTA) and 'Event tree Analysis' (ETA).

FTA tracks down the causes and assesses the hazards that would have been converted to risk leading to the mishap. ETA, on the other hand, traces the resulting consequences and effectiveness of controls.

As such FTA is more a tool for accident investigation, but the process may be reversed to identify hazards in the control implementation system.

A better comparison between FTA and ETA may be as follows:

- Fault trees begin with a top event (E.g. mishap) and work backward toward the initiating events (E.g. causes). It is deductive in nature.

- Event trees begin with an initiating event (E.g. mishap) and work toward the top events (E.g. consequences). It is inductive in nature.

When used quantitatively, these two methods for analysing the factors for causes and consequences of a mishap are more advanced than the ones we have studied so far because they are based on mathematical probability estimates of contributions to and from the mishap and determination of their cumulative effect.

However, the same approach may be used without numbers, to lead to qualitative conclusions.

(b) Principle and Procedure of FTA:

- Fault Tree is prepared by diagramming contributing events to show their relation to each other and to the undesirable event being investigated

- The probability of mishap of each component or of the occurrence of each condition is determined.

- The probability of occurrence of the undesirable event is then determined by calculation from various feasible combinations of the components.

  (a) When any one of two or more basic events is sufficient to cause failure, their combined probability is additive, and is known as an 'OR' operation (or 'gate' in electronic logic).

  (b) When two or more basic events must act together to cause failure, their combined probability is multiplicative, and is known as an 'AND' operation (or 'gate').

- Analysis ends when all the basic causes are found.

- This is known as a 'Top-Down' analysis because we go 'down' (backward) from mishap to causes.

(c) FTA Example:

When rain is anticipated, freshly laid concrete must be protected from it by tarpaulin or other cover, as illustrated in Fig. 19.3. The worst case scenario is if it rains and if there is no or insufficient tarp cover.

Fig. 19.3. - Rain cover for cast concrete

Rain is an "Act of God", but exposure to it may be controlled by the responsible person checking the latest weather report.

Unavailable or insufficient cover material may be due to:

    (i)   Engineer not indenting for it;

    (ii)   Office delaying the order; and/or,

    (iii) Necessary funds not being available.

The possible chain of events leading to the mishap, namely 'concrete affected by rain', is shown in Fig. 19.4.

Likelihood of mishap is found from the computed probabilities of various contributing events.

We shall assume the probability scale as follows:

- 0.1 (1 in 10) for 'frequent'
- 0.01 for 'occasional'
- 0.001 for 'rare', and
- 0.0001 for 'improbable'

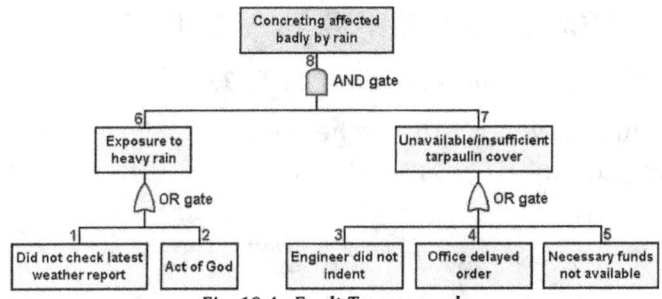

Fig. 19.4 - Fault Tree example

We assign (from experience, statistics, or expert opinion) the following probabilities to various events:

1. Not checking weather report *[→ Rare]* p_1 = 0.001
2. Raining (Act of God) *[→ Occasional]* p_2 = 0.01
3. Engineer not indenting *[→ Improbable]* p_3 = 0.0001
4. Office not ordering *[→ Rare]* p_4 = 0.001
5. Funds not available *[→ Frequent]* p_5 = 0.1

Then, the probabilities at the next level are:

6. Probability of exposure to rain [1 OR 2]:

$$p_6 = p_1 + p_2 = 0.001 + 0.01 = 0.011$$

7. Probability of unavailable/insufficient cover [3, 4, OR 5]: $p_7 = p_3 + p_4 + p_5 = 0.0001 + 0.001 + 0.1 = 0.1011$

8. Probability of adverse effect of rain on concreting [6 AND 7]: $p_8 = p_6.p_7 = 0.011 \times 0.1011 = 0.0011121$

Finally the mishap probability translates to p_8 of 1 in 899. If there are 5 concreting activities going on at the same time in the same environment, then the chances of the hazard being realized is 1 in (899/5), or 1 in 180.

19.8. EVENT TREE ANALYSIS (ETA)

(a) Principle and Procedure of ETA:

Illustrated by the schematic of Fig. 19.5, the procedure for ETA is as follows:

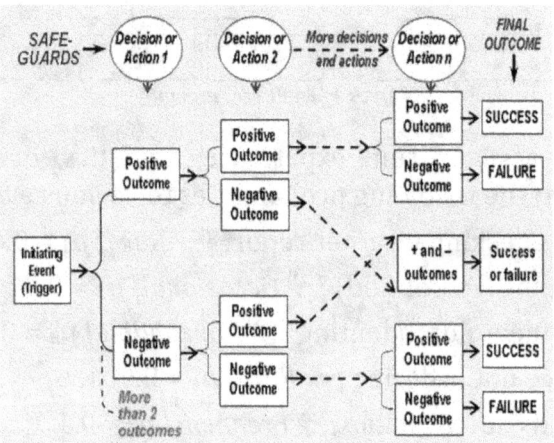

Fig. 19.5. - ETA Schematic

- ETA begins with a mishap, that is, an initiating adverse event or activity ('trigger') such as a component failure, or human error.

- Application of some safeguard ('control') to eliminate or manage the mishap will generally lead to two consequences, referred to as 'outcomes', one positive and the other negative. In practice, there may be more than two outcomes, such as "No flow", "Full Flow", and "Partial Flow". In any case, all outcome probabilities must add up to 1, i.e. 100%.

- Each outcome in turn may confront another situation or safeguard, which will again branch into positive and negative outcomes.

- Each of the various feasible paths is tracked till all the safeguards have been applied, and any intermediate triggers are accounted for.

ETA may be quantified by assignment of a probability of occurrence to each branch at every safeguard.

At the end of this exercise, the list of final outcomes, either 'Success' or 'Failure', or alternatively, final specified outcomes, will be left. Thus the analyst will know which mix of events will be hazardous (and to what extent) and which not.

This is called a "Bottom-up" analysis because we go 'up' (forward) from mishap to consequences.

(b) ETA Example:

We will analyze outcomes of a fire hazard in a building with sprinkler, from malfunction of safeguards.

Analyst has identified and planned for three safeguard failures and their estimated probabilities, as follows:

1. Fire may spread quickly: 'Yes' = 10% ('No' = 90%)
2. Sprinkler system may fail: 'Yes'= 30% ('No' =70%)
3. People inside may not be able to evacuate in time: 'Yes' = 50% ('No' = 50%)

Here, 'Yes' corresponds to failure, and 'No' refers to success. 'Yes' and 'No' must add up to 100%.

Figure 19.6 shows data and its organization format.

When an early hazard cannot be controlled, the next hazard must be faced. Thus, if the fire spreads quickly, the next safeguard is the sprinkler system. If that fails, at least people must escape quickly. If they cannot also escape, consequence will be many fatalities.

INITIATING EVENT	FIRE SPREADS QUICKLY ?	SPRINKLER FAILS ?	PEOPLE CAN- NOT ESCAPE ?	CONSE- QUENCE	SCENARIO No. and %
			0.5 →MANY	MANY FATALITIES	1 0.015 or 1.5%
		0.3 Y	Y N		
	0.1 Y	Y	N 0.5	LOSS & DAMAGE	2 0.015 or 1.5%
FIRE STARTS		N 0.7		FIRE CONTROLLED	3 0.07 or 7.0%
FREQUENCY = 2 / YEAR	N 0.9			FIRE CONTAINED	4 0.9 or 90%

Fig. 19.6. - ETA Example of fire hazard

By Scenario 1, the probability of failure will be the product of the three preceding probabilities, that is,

p_1 = 0.1×0.3×0.5 = 0.015, or 1.5%.

If the fire does NOT spread quickly, people can escape regardless of whether the sprinklers act or not. Hence, the fire may be considered 'contained', with the probability of success for the last Scenario 4 being:

p_4 = 0.9, or 90%.

In Scenario 2, corresponding to people being able to escape ('N' meaning opposite of cannot escape"), but other loss and damage being sustained probability is:

p_2 = 0.1×0.3×0.5 = 0.015, or 1.5% also.

For Scenario 3, with fire spreading quickly, but sprinkler system working, the fire will be controlled, no chance of harm to people; probability of damage will be:

p_3 = 0.1×0.7 = 0.07, or 7%

Scenario 4 depends on the probability that fire will not spread quickly – *that has to be assured!* This needs pre-planning in minimizing total flammable contents in building and other safety measures such as inspection.

Note that the total probabilities of three failures and one success add up to 100%.

As the frequency of fire is given to be 2 per year, probabilities of losses will be as follows:

Scenario 1: Many fatalities = $1/(2\times0.015)$ = 33 years

Scenario 2: Fatalities/damage = $1/(2\times0.015)$=33 years

Scenario 3: Loss and damage = $1/(2\times0.07)$ = 7 years

19.9. 'FISHBONE' OR ISHIKAWA ANALYSIS

This 'Cause and effect' procedure was developed by Japanese professor Kaoru Ishikawa in 1968. As the graphical format for it appears similar to the shape of a fish skeleton, it is called a 'fishbone' analysis.

(a) Principle and Procedure:

Fishbone analysis starts with a mishap, and tracks various feasible causes in different categories.

To facilitate identification and assessment, the causes may be grouped under various categories like in the Canadian model: *"Task–Material-Environment–Personnel–Management"*, or other convenient factors.

In a way, this is a visual organization of the *'What if ...?'* method already discussed. Vegetarians please note: the fishbone analogy is just a reminder of the structure!

There is also a well-known investigative technique called 'The 5-Why" method credited to Toyota, in which one asks *'Why?'* five times sequentially and more often than not, root and contributory causes are revealed.

The fishbone technique follows a similar path until root causes are determined.

(b) Fishbone Example:

Figure 19.7 depicts the analysis to find the causes of why a worker fell from a scaffold. The three categories and two causes for each shown are illustrative and not exhaustive.

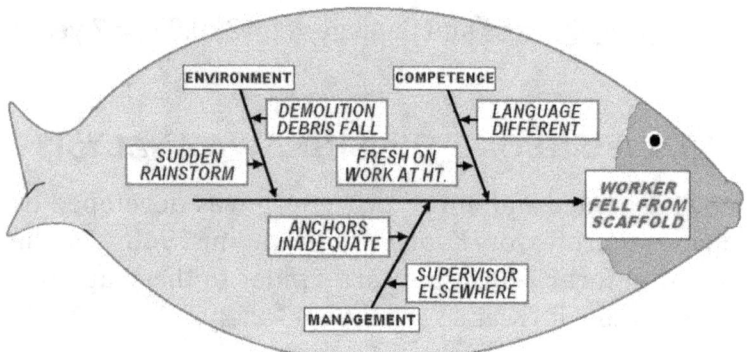

Fig. 19.7. - Fishbone Example

From the findings, one may continue to assess the consequences and recommend suitable controls.

19.10. BOW-TIE OR BUTTERFLY DIAGRAM

As mentioned, this schematic display, depicted in Fig. 19.8, is a way of checking the risk management scheme from causes to mishap to consequences.

The safety route from causes to mishap is via

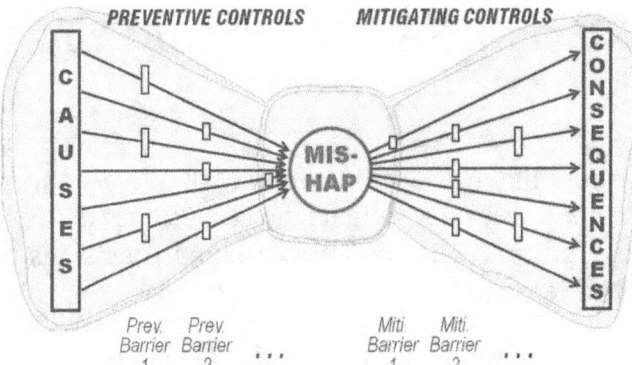

Fig. 19.8. - Bow-tie Diagram schematic

'preventive controls' achieved by barriers (safeguards) against the causes leading to mishap before it happens. Example of a preventive barrier is scaffold guard-rail.

The safety route from mishap to consequences is via 'mitigating controls', achieved by barriers (safeguards) aimed to reduce the impact of the mishap after it occurs. Example of a mitigating barrier is the safety harness.

This will provide a complete status map of the safety and risk management system in place, as well as indicate deficiencies and redundancies.

———

20. THE 5A-WAY TO SAFETY

I would like to close this book with a five-point self-assessment technique which I developed over decades to address my personal and professional decisions, and resolve dilemmas.

It became my '*5A-Way*' to decision-making.

In coming up with this 5A-Way, I *"stand on the shoulders of giants"* (as Newton is alleged to have said) – many all-time great activists and philosophers: such as Mahathma Gandhi, Socrates, and Confucius!

It started with the decision to rationalize and integrate into practical daily life such common adages as *"Think before you leap"*, *"A stitch in time saves nine"*, and *"Who will bell the cat?"*

In due course, I distilled the essence of what the adages basically were saying into two, three, four, and finally five words, finally, at the turn of the millennium, managing to find all five starting with the letter 'A'.

I will frankly admit that here, that my fascination with the five 'digits' found full expression!

I tried the technique on various decision-making situations that crossed my domain, and ended up applying it to workplace safety and risk management – which is the mode I present it here as the *"5A-Way to Safety"*.

The motivation for including this concept in a book on risk management is to remind practitioners that safety management and in particular risk assessment needs to be grown from the ground up, that careful planning and robust implementation are essential, and that constant vigilance and dedication are critical.

20.1. PRINCIPLES OF THE 5A-WAY

Themes of the 5A-Way are presented in Fig. 20.1.

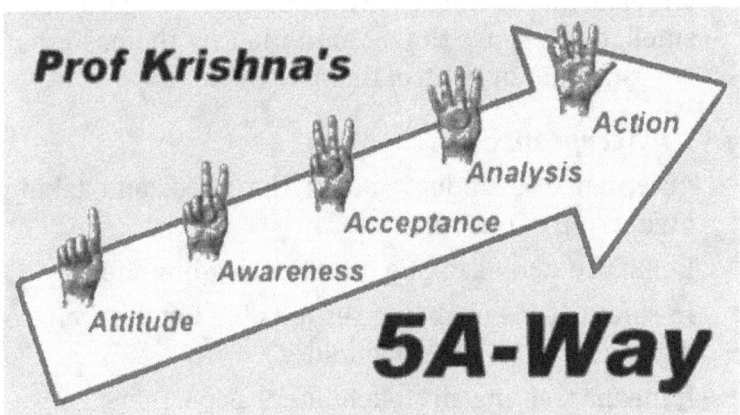

Fig. 20.1. - The 5A-Way

(A-1) Attitude

- Attitude is everything.
- It is the beginning, the foundation for all we want to do.
- In safety, it means that we want all the workers who came in the morning to return home in the evening, safe and sound.
- That we wish to share information and knowledge about incidents and accidents with our cohorts
- And so we can proudly tell ourselves, and also tell others: *"I accept workplace safety as a core value!"*

(A-2) Awareness

- Awareness is knowing what is involved.
- Knowing what is going on around us.
- In safety, it means we must learn what can cause harm, to whom, with what impact.
- It means that we must save our employees from injury or death, to save our property from damage, and our environment and reputation from harm.
- It means that we must see the potential dangers, hear, smell, taste, sense them ... and identify them, so that we can avoid or control them.

(A-3) Acceptance

- Acceptance is understanding the need, and taking ownership of our role.
- Being pro-active, taking initiative, leading the way.
- In safety, it means being responsible for the welfare and safety of all the stakeholders.
- It means treating all stakeholders as partners.
- It means making RM part of our mission and our vision ... not only say the words but also to arrange for funds and personnel to achieve them.

(A-4) Analysis

- Analysis is the key to defining details and making decisions.
- Applying the right theories, using the right tools.
- In safety, it means identifying the hazards, determining if and when accidents may happen, evaluating how bad they may be if they do, and what their combined risk level would entail.
- It means we decide which risks are acceptable, which are unacceptable, and which are tolerable and can be

managed.

- And also what and who can control them, and when.

(A-5) Action

- Action is achievement.
- Putting our money and our effort where our mouth is.
- In safety, it means getting all stakeholders to participate.
- It means documenting the background and all facts, implementing our decisions, re-doing our assessment as required ... getting on with the business of safety, and the safety of our business.
- It entails communicating with all concerned, reviewing progress, continuing to advocate and promote safety first ... safety last.

20.2. APPLICATION OF 5A-WAY TO RISK MANAGEMENT

Almost every report by the Singapore MOM on workplace accident investigations emphasizes the importance of risk assessment in avoiding accidents or mitigating their consequences.

In many instances the lack or inadequacy of detailed or complete risk assessment has been tracked as a contributory (if not primary) cause to the accident.

As a general approach to accident prevention by risk assessment, management may view RA and RM by the 5A-Way as follows:

A1. Attitude:

Workers are the keystone of the success of our plans. Ensuring their workplace safety is to both their and our benefit.

A2. Awareness:

Risk assessment is a leading indicator for accident prevention, and now a pre-requisite to workplace safety. It is often both a legal and professional requirement.

A3. Acceptance:

We accept the responsibility for workplace safety, as well as the commitment to enabling and enforcing it planning for all necessary resources.

A4. Analysis:

We will conduct a thorough risk assessment to analyze every job to determine and rank all hazards. We will recommend all necessary controls.

A5. Action:

We will implement all the recommendations of RA team, consistent with our goals, providing adequate funds and ensuring management involvement. We will empower all concerned right down to the worker.

20.3. APPLICATION OF 5A-WAY TO DENGUE CONTROL

We may look at a specific application.

Singapore has been battling Dengue fever caused by mosquitoes since the 1990-s. In spite of country-wide monitoring and strict enforcement of preventive measures, more than 30,000 dengue cases expected in

Singapore this year (2016).

Considerable publicity is given to the causes of the fever and ways of preventing it. Figure 20.2 shows a poster displayed on metro trains.

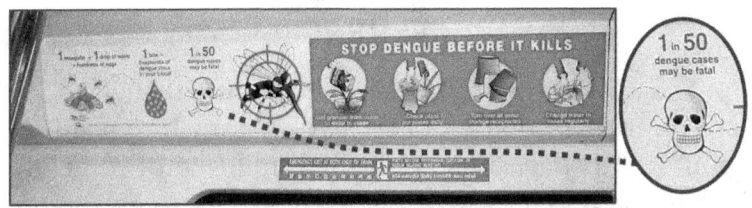

Fig. 20.2. - The 5A-Way applied to Dengue control

The problem of getting the local populace and field work employers involved in this effort may be solved through the 5A-Way as follows:

A1. Attitude: This is in the national interest, including my family and workers.

A2. Awareness: It can kill innocent people, to avoid which concerted action by one and all is necessary.

A3. Acceptance: We have a duty to prevent such harm. I must and will do all I can.

A4. Analysis: Likelihood, severity, and risk are all high.

A5. Action: Check and eliminate stagnant water.

Alright, but how does this affect the workplace?

A local newspaper reported a few years ago: *"PROBLEM SITE – National Environment Agency officers found about 800 mosquito larvae in eleven separate breeding areas at the ABC site in DEF township. A stop-work order was issued from Sept 9 to 15 while the site was cleaned up."*

Delay and thousands of dollars down the drain for

lack of a fraction of the time and money spent or risk assessment and control!

20.4. ONE PERSON CAN MAKE A DIFFERENCE!

The 5A-Way has become second nature to me to such an extent that I don't realize that even my impulsive or spontaneous decisions are rooted in it – until someone brings up the relevance.

Remember the 'hug-rail' example in Chapter 7 (I reproduce Fig. 7.1 here as a quick refresher.)

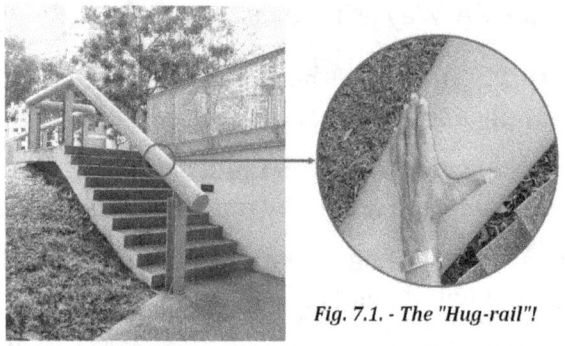

Fig. 7.1. - The "Hug-rail"!

There I used it as an example of risk identification.

It can be used an easy exercise in risk management.
- Hazard: Slipping and falling
- Likelihood (Children, old people etc.) : Medium
- Severity: High
- Risk: High!

But there is more to the story. The children's park is next to my apartment block. The first time I saw the steps, I identified the problem, and pretty soon I started imagining all kinds of nasty harm that might befall innocent children, unwary mothers and old folks.

I took a picture of it and wrote by e-mail to the local political representatives about my fear of accidents at the park and (as a structural engineer) suggesting the simple solution of welding a smaller diameter handrail and vertical bars to the existing pipe, as in Fig. 20.3.

Fig. 20.3. - The 'Hug-Rail' and my recommendation

My friends and colleagues scoffed. But in a week , I got an e-mail reply thanking me for my message and assuring me that the replacement railings *"would be installed on site"* within the month. I checked a few times. When the month went by, I gave up and simply forgot about my impulsive action.

A couple of months later though, I passed by the park again, and what greeted my eyes was not a tack-on patch-up job, but a brand-new stainless steel hand-rail, as in Fig. 20.4!

I am sure you can track my sub-conscious journey through the 5A-Way from 'Attitude' to 'Action'!

When participants in my risk management courses wonder aloud what one member of a team could do to improve an unsafe situation I relate this story and show the pictures, and tell them: ***"You can do it too!"***

Fig. 20.4. - Replacement for the 'Hug-Rail'

CAN one person make a difference?" "Yes indeed!"

20.5. CLOSING THOUGHT

My grandchildren ask me: *"Thatha (– meaning grandpa) ... what are you doing, working so hard, lecturing, writing, visiting construction sites and climbing scaffolds ...?"*

I tell them: *"I am saving lives!"*

Prof Krishna
HAPPY AND SAFE RISK MANAGEMENT
– the 5A-Way, or any other way!

———

ABOUT THE AUTHOR

Dr. N. Krishnamurthy, known as 'Prof Krishna' to his students and colleagues, completes 56 years of teaching and training, research and consultancy at universities, and for Government and private agencies in USA, India, and Singapore.

He received BSc and BE(Civil) degrees from India and MS(CE) and PhD from USA. He taught in two universities in India, three in USA, and three in Singapore. He worked summers and consulted for the Oak Ridge National Lab in USA.

He has to his credit four books on his own and three co-authored, plus over a hundred papers.

Formally trained in structural engineering, Prof Krishna started on computers in the USA in 1959, and has been trying to keep up with applications. He has one book on computer graphics and uses spreadsheet extensively in his study and research.

For the last 15 years, Prof Krishna has focused exclusively on workplace safety and risk management, lecturing, conducting sponsored research, consulting, and publishing on these topics in Singapore, India and USA. He investigates workplace accidents, and appears as expert witness in court cases.

He has one invention on risk management patented in Singapore and Australia, and another related to scaffolding and formwork in the works.

More information on his recent activities may be found from his website: www.profkrishna.com

———

www.ingramcontent.com/pod-product-compliance
Lightning Source LLC
Chambersburg PA
CBHW071837200526
45169CB00020B/1625